Rosamunde Pilcher

# Early Science and the
# First Century of Physics
# at Union College, 1795–1895

## By V. Ennis Pilcher

Published as a Contribution
to Union College's Bicentennial

Schenectady, New York
1994

**Early Science and the
First Century of Physics at Union College
1795–1895**

ISBN 0-9643133-0-8
Library of Congress Catalog Card Number: 94-92379

Printed by Coneco Litho Graphics
Glens Falls, New York 12801

To Edie

# CONTENTS

# ILLUSTRATIONS

# FOREWORD

That science plays an important role in American culture has become a truism in the latter half of the twentieth century. What we tend to forget, however, is that science has played an important role in American culture ever since it gained a firm footing in the curricula of the several colleges established in the colonial and early republican period. During the second quarter of the nineteenth century, dozens of colleges and academies made major investments in science: establishing courses, hiring faculty, and purchasing apparatus that was often sophisticated and expensive. The outlines of this story have been drawn, but the details remain to be fleshed out. Thus the great delight with which I read Ennis Pilcher's account of the first century of physics at Union College.

In recounting this story, Professor Pilcher has raised a number of significant themes. One is that physics was long seen as a part of liberal culture. In the eighteenth century, science was often seen as the handmaiden of religion. Although natural theology long remained an important ideology, it was increasingly supplemented with ideas about mental and moral discipline. As President Smith explained in 1795, by dealing with "nature and her laws," physics "lifts the level of the mind remarkably." Pilcher also discusses the relationship between a liberal physics and a more practical subject such as engineering; the introduction of laboratory practice; and the interplay between local talent and external social forces. His wisest decision, however, was to call attention to books and apparatus. Besides representing substantial financial commitments, these obsolete objects indicate what was taught and how it was taught. In this history, they bring hidden aspects of the past to light, if not to life.

Deborah Jean Warner
Curator, History of Physical Sciences
National Museum of American History
Smithsonian Institution

# ACKNOWLEDGEMENTS

The writing of this book has progressed, on a parallel course, with a project to identify and preserve the many items of early physics apparatus in the Union College Collection.

Thomas Greenslade of Kenyon College, and Deborah Jean Warner of the Smithsonian Institution pointed out to me the historical significance of Union's apparatus and encouraged me to pursue the research which led to this book.

Ellen Fladger, Head of Special Collections, Schaffer Library, provided guidance through the incredible maze of documents in the archives and was an unfailingly friendly source of advice and encouragement. Additional thanks are due to the entire library staff and especially Elizabeth Allen, Archival Specialist. All archival illustrations in this volume are courtesy of Schaffer Library.

Ruth Ann Evans came to the rescue on many occasions with her encyclopedic knowledge of Union College and the history of early Schenectady.

Codman Hislop's monumental biography of Eliphalet Nott was this amateur historian's indispensable guide in all aspects of the work.

I am grateful to several people who took time to read early versions of the manuscript and make useful suggestions. Charles Swartz, Ruth Ann Evans, Carl George, Seyfi Maleki, Ken Schick and Wayne Somers were particularly helpful in this regard.

Special thanks are due to Seyfi Maleki for his generous advice and support and for his many valuable contributions to the apparatus project, to Jack Hogle for his skillful work on the repair and restoration of the early apparatus and to David Peak, whose history of the second century of Physics at Union College has helped round out this study.

Donations from the Union College Physics Department and the Winfred Schwarz Memorial Fund helped to make this publication possible.

My wife, Edie, was a major contributor to every aspect of this publication—research, editing, typing and final preparation of the text.

# INTRODUCTION

As Union College celebrates its bicentennial, it is timely to delve into a long-neglected aspect of its history, as a basis for better understanding the development of this venerable institution.

Existing histories of Union reflect the particular interests and backgrounds of their authors, none of whom were scientists (with the exception of Franklin B. Hough, who wrote a little-known monograph for the federal Bureau of Education). It is not surprising, therefore, that the literature contains few details regarding the scientific curriculum in which Union pioneered and on which rested much of the College's fame in its first century.

It is this aspect of Union's history which has long intrigued me, and which the present work is intended to amplify. Now retired, after teaching physics at Union for 30 years, I have enjoyed following the winding and often tortuous path by which the earliest courses in natural philosophy slowly developed into a modern physics curriculum, while spawning separate specialties in life sciences, natural history, chemistry and engineering.

Fortunately, most of the early records relating to curriculum can still be found in the unusually rich archives of Union's Schaffer Library, so it has been possible to piece the story together through examination of trustees' minutes, catalogs, by-laws, course notes, text books, and lists and descriptions of apparatus.

Comparison with developments at other colleges provides further evidence of Union's leading role in science education during its first hundred years.

## UNION COLLEGE IN 1795.

### Union College in 1795

*Constructed at the corner of Union and Ferry Streets in 1785, it originally housed the Schenectady Academy, and then was used by Union College from 1795 to 1804. After the College moved, it was sold to the City of Schenectady and used for a City Hall and jail. Later Nott reacquired it for his academy; after that it became a cabinetmaker's shop. In 1859 it was extensively rebuilt as a residence for the H. S. Barney Family. Today it is used by the First Presbyterian Church for offices; many of the original architectural features are recognizeable in the present structure.*

# CHAPTER ONE

## UNION'S FORMATIVE YEARS: 1795–1814

A strong emphasis upon science has been a significant feature of Union College since its founding, but the importance of this factor has long been neglected in college histories.

This study is concerned only with the first century of science (particularly physics) at Union; a brief summary of the second century of physics by Professor David Peak is included in Appendix G.

### Union College Founding

Some significant aspects of the College's founding are not widely known, so a brief account is included here, as background for later curricular developments.

Union College evolved from a plan conceived during the American Revolution by members of the Dutch Reformed Church in Schenectady. In 1779 they presented a petition to the New York State Legislature, signed by 973 upstate citizens of diverse religious affiliations, asking that a college be established in Schenectady. No legislative action was taken at that time because of overriding wartime priorities.

In 1784 the Reverend Dirck Romeyn became minister of the Dutch Church and spearheaded a renewed drive to establish a school and academy, in expectation that the latter would soon develop into a college. The academy was founded on February 21, 1785 with Romeyn as president; he divided his time between school supervision and pastoral duties. Over the next several years the academy grew rapidly, but repeated efforts to obtain a charter for "The College of Schenectady" were rejected by the newly established New York State Board of Regents.

The Regents did award a charter for an "Academy of the Town of Schenectady" in January 1793, and its Rules and Regulations were published on March 28, 1793. These indicate that the Academy's four "forms" corresponded exactly to the four years of study offered by established colleges of that time, except for culminating in a degree.

The Rules made a clear distinction between the academy level of studies and the two lower schools—an "English" or grammar school,

1

and a "Latin" or college preparatory school. At that time some 40 students were enrolled at the academy level.[1]

Another petition to charter the Academy as a degree-granting college was rejected in 1794, but at long last the Regents awarded a charter to Union College on February 25, 1795. (The only other college in the state was Columbia, founded in colonial times as Kings College.)

Union was the first college in America to be founded by persons from different denominations, with provisions against control by any one of them. Its name symbolized inclusion of people from all religious and social backgrounds, reflecting the ideals of the federal union. Almost all other colleges at the time were denominational and primarily concerned with ministerial education.

The College's first home was a two-story structure at the northwest corner of Union and Ferry Streets, built in 1785 by the Dutch Reformed Church for their academy. It was only 30 by 50 feet, with two large rooms on each floor, and was soon overcrowded with library, apparatus and classrooms. Students boarded with families in the town.

In the fall of 1795, Union opened with 19 students and one faculty member, Colonel John Taylor, who also served as acting president until the Reverend John Blair Smith arrived in December.

**Colonel John Taylor**

*Served as acting president and was the only faculty member when Union College opened.*

Taylor had originally come to Schenectady in 1792 to head the Schenectady Academy and was an active member of the group which applied for Union's charter. As the only professional educator among the College's founders, he played a major role during its formative years and strongly influenced the design of Union's first curriculum.

A 1770 graduate of Princeton, Taylor had been a staff officer with Washington in several campaigns. After the war, he became Professor of Mathematics and Natural Philosophy at Queens College (later Rutgers) and pioneered in the introduction of science and "practical" studies into the curriculum there.[2]

Union's first three graduates received their degrees in 1797 after attending only two years, a result of the overlapping relationship between the College and the Academy.

## The First Curriculum

Union's first Laws and Regulations, recorded in the minutes of the Trustees' Meeting of December 9, 1795, were very similar to those of the Academy. However, the curricular differences between these two documents reflected the liberal Enlightenment ideals of Union's founders and its first president.

In both programs, the first two years of instruction focused mainly on classical languages and literature; no mathematics was required for admission, but arithmetic was taught after entrance. The last two years were mostly mathematics, surveying, navigation, astronomy and natural philosophy, along with additional studies in Latin and Greek.

This was the standard curriculum of the period, but Union's program broke with tradition in two important ways. (See Appendix A for the full text of the initial curriculum.)

For the first time in an American college, French could be substituted for Greek, both for admission and for all four years of study. Also some traditional classical subjects were omitted to make room for new courses in American History, the American Constitution and history of the different states.

Union's liberal ideas regarding study of modern languages and science were expounded by Union's first president, John Blair Smith. Smith was a former President of Hampden-Sydney College in Virginia, and then Minister of the Pine Street Presbyterian Church of Philadelphia. In his Inaugural Address, he stated:

"...an acquaintance with the French language...is very helpful for increasing our knowledge and getting things done in our travels....

Finally, to the student body as a whole we should open up philosophy, which sometimes is called 'the knowledge of things divine as well as things human.' That portion of this field which is called physics and which, along with astronomy, deals with nature and her laws, lifts the level of the mind remarkably."

According to Samuel Miller, who surveyed American colleges in 1800, the move away from ancient languages was part of a trend toward more utilitarian education:

"...discarding dead languages as the ordinary medium for philosophical publications made such writings more accessible and popular."[3]

## The Natural Philosophy Course

During the earliest years of Union College, all science was included in the natural philosophy course, followed by additional studies in astronomy, and was all taught by a single professor.

Treated primarily as examples of divine revelation of God's handiwork, with little thought of utilitarian value, natural philosophy was a well-established part of the classical curriculum in all colleges and underwent very little change during the first decades of the 19th century.

As scientific knowledge expanded, divisions and then departments evolved. The first distinction was between natural history (i.e. biology and geology) and physical science (from which physics, chemistry and engineering later developed as different fields).

Natural philosophy finally became synonymous with physics during the mid-19th century, and eventually included separate studies in mechanics, heat, light, sound, electricity and magnetism, all considered basic to a general education.

Thus, the history of science at Union and contemporary colleges begins with natural philosophy.

## The First Purchase

Unprecedented financial support for emphasizing instruction in science was provided by the New York State Legislature, two months before the first meeting of Union's Board of Trustees. An Act was passed on April 9, 1795, authorizing:

"...the sum of 1500 pounds, as a free and voluntary gift on the part of the people of this state, to be applied to the purchase of an apparatus of instruments and machines for illustrating lectures in astronomy, geography and Natural Philosophy; and the residue, if any, to be applied to the purchase of such books, for the use of the said college, as the Trustees may think proper."[4]

Receipt of these funds, approximately $7,000, is recorded in the minutes of the meeting of the Board of Trustees on June 22, 1796, as are the arrangements for a major order of apparatus and books, therein designated as "The First Purchase." (Perspective on the generosity of the grant can be gained by comparison: at that time, the President of the College received an annual salary of $750, plus housing.)

The First Purchase included a comprehensive collection of books and a complete set of apparatus for teaching scientific principles. The apparatus was ordered through William Young, a dealer in Philadelphia, from W. & S. Jones, instrument makers in London.

An Invoice of Philosophical Instruments accompanied the shipment, dated Oct. 25, 1797. It listed a fascinating collection of over one hundred items, at a total cost of 201 British pounds. The collection was described by Union's President John Blair Smith as "elegant and valuable" and by Samuel Miller as "respectably large and good."

Miller's 1800 survey of 27 institutions of higher education in America[5] found only 16 operating at the college level.[6] His appraisal of the philosophical apparatus at these 16 colleges placed Union's among the best in the nation.[7]

Instruments in Union's collection provided for demonstrations in all of the topics covered in natural philosophy courses. Among the major items were an eight-inch cylinder electrical machine with medical and philosophical apparatus, a large double barreled air pump, a planetarium or orrery, a brass telescope, a case of "magnetical" apparatus, a Hadley's Quadrant, three microscopes, and a wide variety of items for demonstrations in optics, hydrostatics and mechanics.

Almost all of this early collection was discarded long ago but, fortunately, the accompanying books from the instrument cases are still in the library. Some were written by the instrument makers themselves and provide a clear picture of how physical principles were demonstrated in those very first courses at the College.

Two major pieces from the First Purchase are still in the College's possession and they provide impressive evidence of the fine quality of the instruments and the scope of the collection.

London October the 25th 1797

Invoice of Philosophical Instruments, Ordered by Mr.
William Young, and shipped this day on board the Hazard,
John Drummond, for New York, directed to the care of Dr. Lynn
by W. & S. Jones, Holborn

| | | £ | s | d |
|---|---|---|---|---|
| Case No 1 | An 8 inch Cylinder electrical machine, with a medical and philosophical apparatus, pack'd in a deal Case | 8 | 18 | 6 |
| | A large size double barrelled best finished Air Pump with a raised stage plate, close receiver and gage glasses | 13 | 2 | 6 |
| | A glass ground extra plate for ditto | 1 | 18 | — |
| | A large brass fastening clamp and piece to do. | — | 10 | 6 |
| | A Double transferer with 2 Glass receivers | 3 | — | — |
| | A single do. for a fountain with plate and pipe | — | 18 | — |
| | A long glass receiver for do. | — | 7 | 6 |
| | A 3 fall guinea and feather apparatus, with a long glass receiver for do. | 1 | 18 | — |
| | A pair of middle size brass hemispheres | — | 18 | — |
| | A Gun lock apparatus | — | 18 | — |
| | A set of lead weights and receiver for ditto | 1 | 1 | 6 |
| | An improved set of brass windmills | 1 | 18 | — |
| | A glass to shew the action of the lungs | — | 6 | — |
| | A fountain in vacuo | — | 5 | 6 |
| | A filtring Cup for quicksilver | — | 4 | 6 |
| | A brass plate and piece of wood | — | 4 | 6 |
| | 6 Breaking squares, cage and cap | — | 7 | 6 |
| | A Glass and Brass model of Water pump | 1 | 4 | — |
| | A Tertullian apparatus with receiver | — | 10 | — |
| | An improved bell | — | 10 | 6 |
| | A Balance, beam, and stand | — | 7 | 6 |
| | A large size syringe with lead weight | — | 15 | 6 |
| | A Skin of leather | — | 1 | 6 |
| | A large open swelled glass receiver for pump | — | 14 | — |
| | | 41 | 7 | 6 |

**Page One of the Invoice of Union's First Purchase**

*The invoice of scientific apparatus from W. and S. Jones of London, dated October 25, 1797, was recorded in the Minutes of the Board of Trustees of May 1, 1798.*

Most important is the orrery, or planetarium, a model of the solar system, with a complex system of gears designed to show the relative rates of rotation of planets around the sun, as well as the earth's diurnal rotation, precession of the earth's axis with the seasons and corresponding motions of the moon. By turning a crank and counting orbital revolutions, students gained an understanding of many of the known facts of astronomy. An accompanying text included detailed instructions for its use.

Named for the Fourth Earl of Orrery, who had the first model constructed for him in 1713, this apparatus was greatly esteemed during the eighteenth century. Union's orrery, which cost 21 pounds, was the most expensive item in the First Purchase.

In an unknown sequence of events, Union's orrery was forgotten for a number of years. It had been disassembled and stored, and then overlooked, probably because of changes in the faculty. It was

**The Union College Orrery**

*This major item from the First Purchase is a miniature planetarium, 36 inches in diameter, with planets and moons revolving as they do in the solar system. The planets Neptune and Pluto were still unknown at the time it was constructed. It was designed by George Adams, Instrument Maker to King George III.*

7

# ASTRONOMICAL AND GEOGRAPHICAL
# ESSAYS:

CONTAINING,

I.

A FULL AND COMPREHENSIVE VIEW, ON A NEW PLAN,

OF THE

## General Principles of Astronomy.

II.

THE USE OF THE

CELESTIAL AND TERRESTRIAL

# GLOBES,

*Exemplified in a greater Variety of Problems, than are to be found in any other Work.*

III.

THE DESCRIPTION AND USE

OF THE MOST IMPROVED

## PLANETARIUM, TELLURIAN,

AND

## LUNARIUM.

IV.

AN INTRODUCTION TO

## PRACTICAL ASTRONOMY.

THIRD EDITION.

BY GEORGE ADAMS,

*Mathematical Instrument Maker to His Majesty, and Optician to the Prince of Wales.*

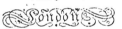

PRINTED BY R. HINDMARSH,

PRINTER TO HIS ROYAL HIGHNESS THE PRINCE OF WALES,

OLD-BAILEY;

AND SOLD BY THE AUTHOR,

NO. 60, FLEET-STREET.

1795.

---

**Title Page of George Adams' *Astronomical and Geographical Essays*, 1795**

*This book, included in the First Purchase, contains complete instructions for use of the orrery.*

rediscovered and restored in 1939 and put on display at the New York World's Fair as part of the Schenectady County exhibit.[8] Since then it has been on permanent display, first in the Physics Building, later in the College's Center for Science and Engineering, then in the lobby of Schaffer Library while awaiting completion of new display facilities in the renovated Nott Memorial.

The other surviving instrument from the First Purchase is a Hadley's Quadrant, an indication of Union's early concern with practical as well as theoretical science. A precursor to the sextant for determining position at sea, it was invented in 1731 by John Hadley. It quickly became the most popular instrument for navigation because of its accuracy and ease of use aboard a moving ship.

Union's quadrant is a handsome instrument of ebony and brass with an ivory scale. It shows signs of heavy usage; navigation was a required subject at the College until about 1830.

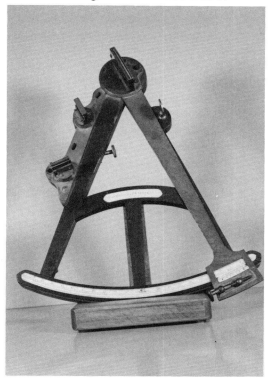

**Hadley's Quadrant**

*This navigational instrument dates from the First Purchase. A precursor to the sextant, it was made by B. Browne of Bristol.*

THE

# DESCRIPTION AND USE

OF THE

## *SEA OCTANT*,

COMMONLY CALLED

# HADLEY's QUADRANT;

WHEREIN

*Its Use is exemplified by proper Examples and Tables,*

AND AN ACCOUNT GIVEN OF THE

## NEW ADDITIONAL APPARATUS

That are applied to it, for determining the Latitude and Longitude
with the greatest possible Accuracy.

TO WHICH IS ANNEXED,

## A DEMONSTRATION OF THE THEORY

OF THIS

## 𝔈𝔵𝔠𝔢𝔩𝔩𝔢𝔫𝔱 𝔍𝔫𝔰𝔱𝔯𝔲𝔪𝔢𝔫𝔱.

THE SECOND EDITION CORRECTED.

---

## BY WILLIAM JONES,

*MATHEMATICAL INSTRUMENT MAKER.*

---

LONDON:

Printed for and sold by W. and S. Jones, No. 135, Holborn. 1795.

[PRICE ONE SHILLING.]

**Title Page of William Jones' Manual on the Use of Hadley's Quadrant, 1795**

## Early Texts and Teachers

Union's first trustees recognized the importance of a good library and they immediately provided for a fine collection of books, applying some of the funds from the New York State grant.

Eight hundred volumes are listed in the 1799 Manuscript Catalog of Union's Library. The collection was of wide breadth and high quality, particularly in scientific works. Miller's 1800 survey showed that Union's library was the largest among those colleges founded after the Revolution, although smaller than collections in the older colonial colleges.

Pre-1800 science books presently retained by the Library are cataloged in a valuable checklist, *Early Scientific Books in Schaffer Library,*[9] compiled by Wayne Somers, an antiquarian book specialist, in 1971. An Introductory Essay in that volume, by science historian Brooke Hindle, provides pertinent background on these books and on early science at Union.

Natural history was well represented in the early library by such works as Oliver Goldsmith's *History of the Earth and Animated Nature,* Buffon's *Histoire Naturel* and William Smellie's *Philosophy of Nature.*

Most of the science taught in American colleges during that early period was taught by men whose primary interest was in other fields, usually the ministry or medicine, and often they were barely able to keep ahead of their students. It was common practice for teachers of natural philosophy to teach mathematics also.

Union was fortunate in that two of its earliest lecturers in natural philosophy, Colonel John Taylor and Cornelius Van den Heuvell, M.D., had unusually strong backgrounds in science.

Although Dr. Van den Heuvell taught natural philosophy for only one year (1798–99), the high standards he set and the topics he covered were unusual for the times. Born and educated in Holland, he had served on the faculty of Leyden University, left for France in 1787, then emigrated to New Jersey where he spent a year before moving to Schenectady. Here, he practiced medicine with Dr. Dirck Van Ingen, Union College Treasurer, and they operated an apothecary two doors north of the Dutch Reformed Church.[10]

A student's notebook from Van den Heuvell's course, 145 pages long, provides a clear picture of the science instruction given to Union seniors. Although the author is not identified, it may have been Hubbell Loomis, an early protégé of Eliphalet Nott, whose reminiscences about Dr. Van den Heuvell were published in *Union College Magazine* in 1871.[11]

Entitled "The Substance of a Course of Lectures on Natural Philosophy," the notebook contains comprehensive and well-organized lecture notes, enlivened by descriptions of demonstrations, rather than the usual question and answer format. The first eight pages are devoted to chemistry, followed by sections on statics, dynamics, pneumatics, hydrostatics, hydrodynamics, heat, light and sound. A large number of demonstrations are described (including thirty-two on the properties of air), which indicates that Van den Heuvell made good use of the College's newly-acquired apparatus.

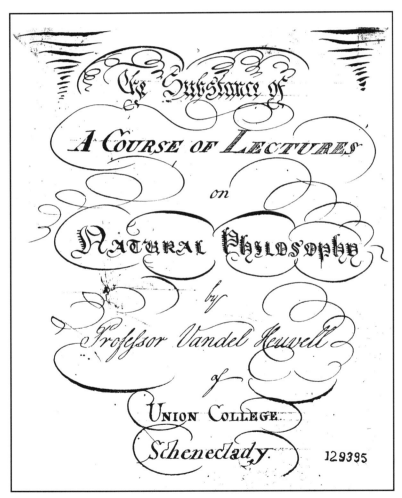

**Title Page of Student's Notebook from
Professor Van den Heuvell's Course, 1798**

These notes contain frequent references to five British texts, four of which the College still owns. The most frequent references were to William Nicholson's *Introduction to Natural Philosophy,* an American edition of a popular text first published in London in 1783. This rather elementary two-volume survey, written for "those who know very little mathematics" was typical of the group. It contained practically no algebra, depending almost entirely upon simple geometric arguments, and reflected the low level of mathematics in colleges at that time.

Van den Heuvell's lectures reflected his continental education, supplementing the English texts with more modern concepts. For example, he discussed the possibility that light was a wave-motion, a theory rejected out-of-hand by English authors, all of whom supported Newton's corpuscular theory of light.

Dr. Van den Heuvell also taught astronomy that year and notes from this course also survive, dated March 15, 1799. These are in a less innovative form containing questions and answers, with references mostly to Nicholson's text. Astronomy courses of that day utilized very few calculations and consisted mainly of definitions; this one conformed to that pattern.

Education at Union suffered a setback when both science teachers were lost within a short period. Dr. Van den Heuvell died in April 1799, at age 39; John Taylor contracted yellow fever on a trip to New York City and died November 5, 1801, at age 49.

In the fall of 1801, Benjamin Allen was appointed Professor of Mathematics and Natural Philosophy; a harsh disciplinarian, he proved to be an extremely unpopular teacher, although he remained for nine years.

During this period, the college was also undergoing rapid turnover in its top leadership. President Smith had resigned in 1799 and returned to his pastorate in Philadelphia. He was succeeded by Jonathan Edwards Jr., son and namesake of the distinguished theologian, who assumed the presidency of Union in May 1799, but died only two years later, in August 1801.

Reverend Jonathan Maxcy succeeded Edwards in 1802. Maxcy had been President of Rhode Island College, the same institution from which Professor Allen had come two years previously. Records indicate that both men were more concerned with discipline than with education.[12]

A major rewriting of the College Laws was undertaken by Maxcy in 1802. These new regulations adopted, almost word-for-word, an ultra-strict system of discipline laid down in the Laws of Rhode

13

Island College. It should be noted that most college students at that time were between the ages of 14 and 18, and their boyish enthusiasms required stern but judicious handling. When severe laws were harshly enforced, unhappy results followed.

The Laws also defined some curricular changes, reverting to a more traditional program. Study of French language and literature was decreased, and courses in American history were replaced with Ancient and Modern History. Science and mathematics were still concentrated in the junior and senior years; the amount of mathematics was increased.

Some highly restrictive regulations applied to faculty as well as students:

> "Each Professor shall deliver to the students, previous to each lecture, questions, the answers to which shall involve the principal points treated of in said lecture; and the students are required during the delivery of the lecture to collect from it and write down answers to said questions. These shall be exhibited to the Professor."

This regulation was continued in College Laws for about 50 years, but records do not show how widely it was observed.

Jonathan Maxcy resigned in 1804, after two turbulent years in which students often protested against his harsh rules; he then became President of South Carolina College.

## Nott Begins a Long Presidency

Reverend Eliphalet Nott accepted appointment as Union's fourth president in 1804 and remained in that capacity until 1866, spanning a period of vast changes in the country as well as the College. With an M.A. from Rhode Island College, Nott had been Principal of Plainfield Academy in Connecticut, then pastor and teacher in Cherry Valley, New York and later Minister of the First Presbyterian Church of Albany. He had been a member of Union's Board of Trustees since 1800. Although only 31 years old at the time he took office, Nott was already known as a master of oratory, a profound thinker, a man of broad interests and abilities and an inspiring leader.

When Nott became president, the young college was suffering from serious financial difficulties as well as low morale, a result of the rapid turnover of presidents and the "prison" atmosphere created by the previous administration. Change, however, was already in the air. The enrollment was increasing rapidly and a new and much larger building for the College was nearing completion one block eastward.

Originally called "College Hall," later known as "Stone College" and then "West College," the handsome four-story edifice contained a chapel, library and philosophical apparatus, natural history collections, classrooms, dormitories and an apartment for the president's family; additional dormitories were built on North College Street. Two lower rooms of the building housed an academic school connected with the College, which Nott had established shortly after taking office.[13] The seventh form in this school constituted the freshman class of Union College.

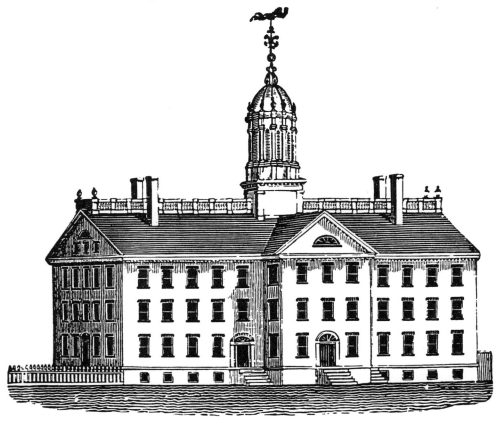

**College Hall, also called Stone College,**
**was later known as West College**

*Designed by Philip Hooker, it was located on Union Street between College Street and today's Erie Boulevard. This building housed Union College between 1804 and 1814, was then sold to the City of Schenectady for use as a City Hall, and was reacquired by the College in 1831 to house instructional facilities for freshmen and sophomores. It was rebought by Schenectady in 1854 and used as a school until razed in 1890.*

Nott's first baccalaureate address in 1805 reflected his belief in the philosophy of the Enlightenment, which had been characterized by Immanuel Kant as "Dare to know." Speaking to a large audience in the old Dutch Reformed Church, he sought to inspire his 13 graduates:

> "Go...excel in whatever is splendid, magnanimous and great; with Newton, span the heavens, and number, and measure the orbs which decorate them—with Locke, analyze the human mind—with Boyle, examine the regions of organic nature...ponder the mysteries of Infinite Wisdom and trace the Everlasting in his word, and in his works."

Nott eagerly embraced the idea of an orderly cosmos as described by Isaac Newton, and he was thrilled to observe the Great Eclipse of 1806 with the aid of the College telescope. His careful records of that event are in the College Archives.

By 1807 Nott was in firm control of College affairs and he chose this time to make significant changes in the curriculum, conforming to a resurgence of political conservatism which was rapidly replacing the liberal ideas of the founding fathers.

Language studies were always an indicator of such changes; Greek resumed its classical importance, and French could be studied only "...under such regulations as the President may prescribe." All courses in modern history were dropped and replaced by additional courses in classics.

More significant, however, than these bows to anti-Enlightenment prejudices, was an increased emphasis upon science. The revised curriculum of 1807 reflected Nott's belief that scientific knowledge was basic and essential in education, and that it would play an increasingly important part in the mainstream of American life.

In addition to placing more emphasis on astronomy and natural philosophy, and mathematics as the language of science, the new program added a separate course in "chymistry" for seniors—one of the first such courses in America to be taught outside a medical school.

Chemistry, with its growing importance in agriculture and medicine, represented the first major step at Union toward the new "practical" science which Nott believed would soon change the world. In 1809 he persuaded a young faculty member, Thomas Brownell (Union, 1804) to change his field of instruction from Greek to chemistry. Brownell then spent a year in Europe studying and purchasing chemical apparatus. Upon his return, he became Union's first Professor of Chemistry. (A history of chemistry at Union College was written by the late Professor Egbert Bacon, whose unpublished manuscript is in the College Archives.)

Brownell also taught rhetoric and mineralogy; he presented his collection of 2,000 minerals to the College when he resigned in 1818 to enter the Episcopal ministry. He later became Episcopal Bishop of Connecticut and was a founder and first President of Trinity College. He was the first of many Union graduates to become a college president.

Science teaching at Union was still hampered, however, by Benjamin Allen's unpopularity; finally in 1809, student riots over disciplinary measures led to his forced resignation.

Allen was replaced by Ferdinand Rudolf Hassler, an outstanding Swiss mathematician and astronomer, and one of the first European-trained scientists to teach full-time in an American college.

Eliphalet Nott was very pleased with this appointment, writing to his brother, Samuel, in Connecticut on April 14, 1810, that "...Hassler goes through a course of lectures in the manner of the most approved European professors." Nott was also impressed with Hassler's large library and collection of apparatus, and perhaps also by his practical approach to science. A student of that year reported that Hassler "...had a workshop in the upper part of his house where with his own hands he fabricated many things pertaining to his scientific pursuits."[14] At Hassler's request, the college placed a sizeable order for new scientific apparatus with John Prince, a dealer in Salem, Massachusetts.

Unfortunately, Hassler was not an effective teacher, hardly caring or noticing that students would occasionally vacate the classroom during particularly complex demonstrations. He taught at Union for only one year, then left to become the first head of the U.S. Coast Survey. He remained Nott's lifelong friend, however, and undoubtedly expanded Nott's horizons with regard to science. Nott encouraged Hassler to write two mathematics textbooks which were used later at Union.[15]

The next Professor of Natural Philosophy was Thomas Macauley, who graduated from Union in 1804 and then served five years on the faculty as a Latin teacher. Macauley was

**Ferdinand Rudolf Hassler**

**Professor of Natural Philosophy, 1810–11**

*Hassler was a major influence upon President Nott in regard to science education.*

another professor who was induced by Nott to change his field of instruction. At this time there were 118 students in the College and the faculty consisted of the president, four professors and three tutors.

Macauley took his new assignment very seriously and made important additions to the philosophical apparatus through purchases and donations and by constructing some items himself. Additions for which he was responsible include a large telescope, valued at $312 (a comparable sum in 1994 would be roughly $2,500), a small telescope, a theodolite, a very rare Gunter's Quadrant and a variety of electrical apparatus valued at $493.

Macauley used Enfield's *Institutes of Natural Philosophy*[16] as his primary text; he probably had been introduced to it as a student under Professor Allen, and continued to teach from it for many years thereafter. The Enfield text was used at most American colleges during that same period:

> "...from 1790 to 1830, almost without exception, college students in America had their first and last look at physics through recitations from Enfield's text."[17]

The universality of this Enfield text clearly indicates the low level of science education in America at that time. Published in London in 1783, it was a compendium of material copied from much older books. The book followed the format of Newton's *Principia,* with propositions and scholia and tedious geometric arguments and was at least fifty years out of date. The astronomy section contained many errors, never corrected in later editions, including a statement that the sun and moon were inhabited.

Union's library list of 1815 shows thirty-three copies of this text and the present library has copies of three different editions, two of which are well worn and contain student annotations reflecting much use over the years. One bears the following enigmatic inscription:

> "Oh! Enfield, thou art the joy of my life...
> I meditate upon thee by day and dream of thee
> by night—surely we are congenial spirits."

## The New Campus

In 1814 Union College began to move from overcrowded Stone College to its present Ramée campus, the first planned campus in America. At that time the new campus was a pastoral site on the outskirts of Schenectady, a pleasant and quiet hill above the city. The philosophical apparatus and lecture rooms were soon relocated in spacious quarters in North Colonnade (the eastward extension of North College).

The academic school, which had been in Stone College since 1804, was merged with a revived Schenectady Academy which Nott re-established, under direction of a Union faculty member, in the old Academy building at the corner of Union and Ferry streets. Union's freshman class remained downtown as part of the Academy.

President Nott began experiments in "caloric," seeking to design a wood-burning stove for heating the rooms in North and South Colleges. These efforts produced a series of stoves; the first was known as "the coffin" and models of it were soon installed throughout the new buildings.

Nott's experiments continued for many years, resulting in 30 patents and culminating in the first successful stove in America to burn anthracite coal. He then established a factory in Troy to manufacture his "Nott Stove," and it was so popular that his name became a household word before the Civil War. In addition to turning a handsome profit, this work established him as a leader in early experimental science and technology.[18]

**Ramée Design for the Union College Campus**

*Engraving by J. Klein, circa 1821*

# CHAPTER 2

## DEVELOPING SCIENTIFIC LEADERSHIP: 1815–1826

After Union moved to its new campus, the number of students increased rapidly. There had been 13 graduates at Nott's first commencement in 1805, there were 38 in 1815 and 50 in 1816. These numbers were similar to those at Harvard and Yale, but by 1824 Union's graduating class of 76 was the largest in America.

As Union entered its third decade, new influences affected the curriculum. The country was expanding rapidly, and there was need for men with practical knowledge to undertake explorations, construct roads and canals and develop water power for industry. Union College was particularly sensitive to these needs because of its location, close to New York's frontier and to the Erie Canal then under construction. It was obvious that trade, settlement and development would follow the expanding frontier, opening many exciting economic and professional opportunities to graduates.

At that time few American colleges offered any practical courses and there was growing criticism that classical studies were irrelevant to the needs of the "active" man and the "spirit of the age."[1] Despite dramatic changes taking place outside the colleges, most curricula, textbooks and teaching methods had changed little since colonial times. Courses in mathematics and natural philosophy had become obsolete because they failed to include important developments taking place in continental Europe. Among these were the transformation by LaGrange and Laplace of Newtonian mechanics into a more useful mathematical form, and a number of fundamental discoveries in physics including those on the nature of light by Young and Fresnel and on the nature of electric current by Volta and others.

### The First "Parallel" Curriculum

In 1815, in an effort to meet demand for a more practical education, Union published new College Laws. In addition to the classical curriculum of 1807, these Laws contained an alternative curriculum which replaced some Latin and Greek studies with more chemistry and natural philosophy and a course in fluxions (differential calculus). For the first time at Union there was a separate course listing for natural history, using *The Philosophy of Natural History* by William Smellie. Although it is not clear just how seriously the new

21

curriculum was pursued, it marks a beginning of Nott's curricular experiments which finally resulted in the tradition-shattering Scientific Course of 1828.

## New Science Books

A new library catalog, issued in 1815, shows that holdings in science and mathematics had grown from 35 to 111 titles since 1799; many of these are still in the current collection. There were multiple copies of some texts, as students were not expected to purchase personal copies. The collection displayed sophistication as well as variety; it included a history of mathematics and a rare 1544 edition of the *Works of Archimedes,* still in the present collection, as well as seven journals of learned societies including *Transactions of the Royal Society, Memoirs of the American Academy, Memoirs of the French Academy,* and eleven volumes of *Philosophical Magazine.*

The growing importance of books and apparatus was recognized by the trustees in 1816, at which time they established a standing committee to "...examine the library, philosophical and chemical apparatus and report annually to the board." This was one of the first such committees to be established in an American college. Significant additions to the library in 1820 included works by Priestly, Boyle and a five-volume set of *Newton's Works.*

President Nott reported to the Board of Trustees in 1823 that Union's chemical and philosophical apparatus was "inferior to none in America."[2] The College Treasurer's Report of 1824 provided further evidence of high quality, placing a valuation upon the classical library—heart of the traditional curriculum—of $5,648.27, and upon the philosophical library and apparatus of $14,114.82.[3]

A current of change was also rippling through American colleges because of changing patterns in secondary education. The number of academies had increased dramatically, producing graduates who were older (usually 17 or 18) and better prepared than ever before. Many academies were providing both scientific and classical instruction equivalent to that of some colleges,[4] so that colleges perceived a need to upgrade their courses or adjust to declining enrollments.

In response to these challenges, representatives from Union, Bowdoin, Middlebury, Vermont, Harvard and Yale met in New Haven in September 1818 to discuss the possibility of adopting uniform admission requirements. Texts and other curricula matters were also discussed. These common interests led to the founding of The Collegiate Convention in Boston in 1819 by a group of professors from the same institutions plus Brown, Dartmouth and other colleges. Thus began a movement toward collegiate cooperation on admission and curricular standards and, simultaneously, competition for students.[5]

Shortly thereafter, colleges began publishing catalogs which contained details of their curricula, where formerly they had only contained lists of faculty and students.

During the 1820s major improvements were made in mathematics and natural philosophy texts used by the leading colleges. By 1828 Union had added some 35 or 40 recent French texts to its library and these transformed the teaching of mathematics and natural philosophy. Because of their studies in French, Union's science students were easily able to read these books in the original editions. Replacing Enfield, they included works by Lagrange, Legendre, Biot, Laplace and Boucharlat covering recent discoveries in optics, sound, electricity and electromagnetism in a modern mathematical framework. The library still retains many of these historically important volumes.

## Professor Macauley and Gunter's Quadrant

Old books have survived better than old laboratory apparatus. Remarkably, however, a very early and valuable instrument, which once belonged to Professor Macauley, has recently been discovered on campus. College Archives have yielded some interesting information about it.

Professor Macauley had been very ill and unable to work for most of the academic year 1817-18. As a tangible expression of his gratitude to the College for continuing to support him during that period, he wrote a letter to the trustees on July 21, 1818, announcing a gift of valuable apparatus—five mathematical instruments—which he asked the Board to accept:

> "...as a small acknowledgement for the indulgence extended to him by your honorable Board, while he was sinking under the influence of disease."

It had long been assumed that all of these instruments were lost or discarded many years ago, but one of them surfaced just recently. It was among a boxful of unidentified, tarnished brass pieces which had been saved by the late Professor William Stone, when Carnegie Engineering Building was remodelled in the 1980s. I was asked to look over the box and keep anything of interest, and was thrilled to discover a very valuable and venerable instrument known as Gunter's Quadrant.

**Gunter's Quadrant**

*Dating from the mid-17th century, this is the oldest instrument in the Union College Collection. Originally it belonged to the renowned General Monk, who served on both sides during the English Civil War.*

Invented in 1618 by Edmund Gunter of Gresham College, London, it is described in detail in a 1673 edition of his works, a copy of which is in Schaffer Library. Easily portable (about six inches in diameter), it was used for telling time by the sun and stars, as well as for measurement of azimuth and elevation.

In his 1818 letter to the trustees, Professor Macauley described it as "very rare...formerly the property of General Monk, in the reign of Charles Second." It probably was used in the English Civil War during the mid-17th century, possibly as an aid to artillery, and is the oldest instrument now owned by Union College.

After the Restoration, General Monk became the first Duke of Albemarle. Macauley stated that he had acquired the quadrant from a lineal descendant of General Monk, but did not indicate how and when it came into his possession; he may have procured it when visiting his native Ireland in 1815 to receive an honorary degree from Trinity College, Dublin.

Thomas Macauley resigned to enter the ministry in 1822, after sixteen years of service to the College, and later became the first President of Union Theological Seminary.

*Jennies*

# THE
# WORKS
OF
## *EDMUND GUNTER:*

Containing the *Description* and *Use* of the

## Sector, Cross-staff, Bow, Quadrant,

And other Instruments.

With a Canon of Artificial Sines and Tangents to a
Radius of 10.00000 parts, and the Logarithms
from an Unite to 10000 :

The Uses whereof are illustrated in the Practice of

*Arithmetick,* } } *Astronomy,* } } *Dialling,* and
*Geometry,* } { *Navigation,* } { *Fortification.*

And some Questions in Navigation added by Mr. *Henry Bond,* Teacher of
Mathematicks in *Ratcliff,* near *London.*

To which is added,

'The Description and Use of another Sector and Quadrant,
both of them invented by Mr. *Sam. Foster,* Late Professor of Astronomy
in *Gresham* Colledge, *London,* furnished with more Lines, and differing
from those of Mr. *Gunters* both in form and manner of Working.

## The Fifth Edition,

Diligently Corrected, and divers necessary Things and Matters ( pertinent
thereunto ) added, throughout the whole work, not before Printed.

## By *William Leybourn,* Philomath.

LONDON,
Printed by A. C. for *Francis Eglesfield* at the *Marigold* in
St. *Pauls* Church-yard. MDCLXXIII.

Title Page of the 1673 edition of *The Works of Edmund Gunter*

25

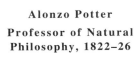

Alonzo Potter
Professor of Natural
Philosophy, 1822–26

Benjamin Joslin, M.D.
Professor of Natural Philosophy,
1827–37

## Further Changes in Faculty

Although most professors of natural philosophy at Union in the first quarter of the nineteenth century were primarily interested in the ministry (a situation common to most colleges at that time), they struggled valiantly to raise the level of instruction in science.

Alonzo Potter (Union, 1818), Eliphalet Nott's son-in-law, was Tutor and then Professor of Natural Philosophy from 1819 to 1826, during which time he published a text on descriptive geometry and pursued theological studies. He made several major additions to the apparatus, including a magnetic machine, a machine for decomposing water, a steamboiler for high pressure, Wollaston's Goniometer and six forms of crystals.

Potter left to serve as pastor of St. Paul's Episcopal Church in Boston, then returned to Union in 1831 as Professor of Moral and Intellectual Philosophy and also became Vice President of the College in 1838. In 1840 he published *The Principles of Science, Applied to the Domestic and Mechanic Arts,* which was used in "technology" courses at Union and was revised in 1860. Alonzo Potter resigned in 1845 to become Bishop of the Episcopal Diocese of Pennsylvania, a post which he held until his death.

Francis Wayland (Union, 1813) was another who vacillated between the ministry and college teaching. He served as a tutor at Union from 1817 to 1821, then entered the ministry, but returned to replace Potter in 1826 as Professor of Natural Philosophy. He remained only one year before being appointed President of Brown University, where he pioneered in introducing many of Eliphalet Nott's curricular ideas stressing scientific and practical studies.

Dr. Benjamin Joslin (Union, 1821) succeeded Wayland in 1827 and served until 1837. He was the first Professor of Natural Philosophy since Rudolf Hassler whose primary interest and training was in science and he raised the level of scientific instruction.

He had received his M.D. in 1826 and practiced medicine while teaching at Union; his interests and accomplishments were diverse. He was author of two books on homeopathic medicine, which were republished in Europe, and many articles on observations and theories in physics, meteorology, and medical science. He edited the American edition of Lardner's *A Treatise on Hydrostatics and Pneumatics* (1832), published his own *Meteorological Observations and Essays* (1836) and made suggestions for improvements in air pumps.

He also assisted Stephen Alexander (Union, 1824) in observing star occultations. Alexander, who had a long and distinguished career as Professor of Astronomy at Princeton, was one of Union's earliest graduates to become an outstanding scientist.[6]

Joel Nott (Union, 1817), son of President Nott, served as Tutor and then Professor of Chemistry and Mineralogy from 1820 to 1831. During this period he studied in Europe for a year and traveled to the Indian country of Michigan to "make mineralogical examinations and a geological survey."[7] It was customary at that time for the same professor to teach chemistry and natural history, which included mineralogy, geology and botany.

## Union Attains Top Ranking

Under Eliphalet Nott's direction, and with a superior faculty and a period of relative prosperity, Union grew in size and reputation, attaining a position (with Harvard and Yale) as one of the three best and largest colleges in the country.

The flexible curriculum attracted many students, including transfers from other colleges and numerous "partial" students who attended only the scientific courses, resulting in particularly large science enrollments.

Always looking ahead, Nott was actively involved at the forefront of technological change, and was also inventive in finding financial support for Union. Since 1813 the New York State Literature Lottery had yielded a steady income for the College; Nott was personally reaping profits from the manufacture of his anthracite-burning stoves. In 1826 he began experiments which led to his patented "tubular boilers" for steamboats, demonstrated on the historic run of his *S.S. Novelty* from New York to Albany. Although it was not a commercial success, this was the first Hudson River liner to be powered solely by anthracite coal.

After a decade of innovation, Nott was ready to introduce another new curriculum, and a new era in scientific and technical education.

**Student Room with Nott's "Coffin" Stove, *circa* 1870**

*Nott designed these simple, wood-burning stoves specifically for heating student rooms. They proved to be almost indestructible.*

### "Men of Progress" by Christian Schussele

*(Courtesy of the National Portrait Gallery, Smithsonian Institution)*

*Eliphalet Nott is shown seated in the center of a group of 19 notable inventors of his age "who had altered the course of contemporary civilization." The painter was commissioned to portray a grouping which actually never happened; between 1857 and 1862 he visited each sitter's home to make sketches from life. Left to right, men and their inventions:*

1. *Dr. William T. G. Morton  (ether as an anesthetic)*
2. *James Bogardus  (cast-iron fireproof building)*
3. *Samuel Colt  (revolver)*
4. *Cyrus McCormick  (reaper)*
5. *Joseph Saxton  (fountain pen & "ever-pointed" pencil)*
6. *Charles Goodyear  (vulcanization of rubber)*
7. *Peter Cooper  (first locomotive in U.S.)*
8. *Jordan L. Mott  (coal-burning cookstove)*
9. *Joseph Henry  (experiments in electricity, first Secretary of the Smithsonian)*
10. *Eliphalet Nott  (thirty patents on stoves and boilers)*
11. *John Ericsson  (first screw-propelled man-of-war)*
12. *Frederick E. Sickels  (first cut-off for steam engines)*
13. *Samuel F.B. Morse  (telegraph)*
14. *Henry Burden  (horsehoe machine & "rotary concentric squeezer" for rolling puddled iron into bars)*
15. *Richard Hoe  (cylinder press and type-revolving press)*
16. *Erastus B. Bigelow  (power loom for carpets)*
17. *Isaiah Jennings  (friction match, threshing machine, etc.)*
18. *Thomas Blanchard  (tack-making machine & multipurpose lathe)*
19. *Elias Howe  (sewing machine)*

# CHAPTER THREE

## UNION COLLEGE'S HEYDAY: 1827–1860

The development of science education at Union College was the major reason Union attained preeminence during the first half of the 19th century. Through a fortunate combination of circumstances, Union succeeded where other colleges failed, in responding to the challenges of a more technical age.

During the 1820s and 1830s many colleges were groping for new ways to serve the educational needs of the times and to attract more students, but Union was the only one to achieve a viable alternative to traditional studies.

The University of Virginia, under Thomas Jefferson's guidance, established a short-lived liberal curriculum in 1825 which was decades ahead of its time. It offered a choice of eight parallel programs, including one in Mathematics and Natural Philosophy. The experiment had to be abandoned, however, because enrollments thereafter dropped precipitously.[1]

Harvard also adopted a statute in 1825 which established several departments of instruction and allowed students to specialize to a limited degree. However, Harvard would not grant a degree to students who specialized in scientific or technical subjects, thus dooming the program from the start. In the decade from 1829 to 1839, Harvard's enrollment decreased from 247 to 216, while Union's increased from 218 to 315.[2]

In 1826 Amherst created a new parallel curriculum which substituted modern for ancient languages and introduced modern history, as well as additional courses in science and mathematics. Here too, however, the Bachelor of Arts degree was not awarded for these studies and the program was soon dropped.[3]

Similar reforms were introduced at about this time at Columbia, Bowdoin, the University of Vermont, and the University of North Carolina.[4]

All of these programs failed because they were pursued half-heartedly; many faculty and administrators considered them a threat to traditional curricula. In 1828, the same year in which Union was to announce its new Scientific Course, Yale published a report which clearly stated the opposing conservative position. It stoutly defended traditional classical studies against incursions of modern languages and new sciences. Clearly, Union's outlook differed from those at the older colonial colleges.

## Union's Unique Curriculum—The B.A. in Science

At Union in 1828, the time was ripe for the introduction of a scientific curriculum which was equal in every way to the classical curriculum and which also culminated in the B.A. degree.

This was described by Codman Hislop, Nott's biographer, as

> "...breaching the great wall of the then traditional American college curriculum."[5]

Hislop devoted much of his life to a biography of Eliphalet Nott, and he carefully documented the progress of Nott's thinking in regard to science education.

Union's new program was spectacularly successful. Largely as a result of Nott's leadership, it flourished at a time when similar curricula at other colleges were failing. Several reasons for Union's success were identified by Professor Frederick Rudolph of Williams College in his definitive study of the American undergraduate curriculum. He observed that Nott had long supported significant purchases of scientific apparatus, had carefully selected superior professors, and had, himself, personified the "optimistic materialism" of the age. At Union, he said, Nott proposed to lead:

> "...a great Christian-scientific assault on nature, an assertion of man's dominion over nature through science in order to achieve moral and material perfection."[6]

> "...The happiest curricular arrangements of the first half of the nineteenth century were made under Eliphalet Nott at Union College....Union was the best of all possible worlds: it believed in the classics and it believed in the new subjects; it believed in a sound moral education and it believed in the application of science to the conquest of the American continent. It believed in young men: it welcomed the Greek-letter fraternity movement to its campus and it had no qualms about welcoming as transfer students young men who had experienced disciplinary collisions at their colleges of origin. Union believed in itself and perhaps that was all that was necessary to make a vital difference—that and the driving presence of Eliphalet Nott..."[7]

The essence of Union's plan was described in the Trustee's Minutes for July 24, 1827, in the following resolution:

> "Resolved, that the faculty be authorized to arrange the studies in this institution as far as practicable in such manner as to afford a choice between the ancient and modern languages and also between branches abstract and sci-

entific and branches practical and particular and report the same to this board."

Union's 1828 catalog broke new ground in two ways: first, by announcing that the new Scientific Course would lead to the B.A. degree, and second, by stating that students were free to choose any combination of the Classical and Scientific Courses which their "taste or contemplated business in life dictated"[8] thus creating what was probably the first truly elective system in an American college. (See Appendix B and Appendix C to compare the curricula for 1827 and 1828.)

Nearly a third of the student body was soon enrolled in the Scientific Course, which was directed toward professional preparation for such fields as engineering, education, medicine, law and mining. It was such an attractive alternative to the standard classical offerings that many students who had little interest in science elected the course, to receive a broader education.

Most students entering Union in the late 1820s and the 1830s came from Schenectady Academy or Albany Academy, which Nott had helped found in 1813, and they had much better preparation than in earlier years. According to the 1828 Union College catalog:

"The Freshman Class, for the most part, constitutes a department in the [Schenectady] Academy, and is taught by the Principal thereof."

At that time, the principal of the Schenectady Academy was also a faculty member of Union College.

Transfer students from other colleges into Union became so numerous that Union's senior classes increased dramatically in size. In 1830, only two years after the Scientific Course was introduced, Union graduated 96 students, Yale 71, Harvard 48 and Princeton 20.[9]

So many Union graduates entered the teaching profession and became top administrators that Union became known as "the Mother of College Presidents." By 1845, 30 former students had become college presidents, including Thomas Brownell who opened Trinity in 1823, and Francis Wayland who became President of Brown in 1827, Leonard Woods of Bowdoin, Silas Totten of Trinity, Henry Tappan of the University of Michigan and John Raymond who was President successively of Madison, Rochester and Vassar. Most of them were exponents of Nott's curricular reforms, with emphasis upon science and upon free electives. It was through the zealous efforts of such men that, according to Hislop, Nott's radical curriculum of 1828 became common in America by the time of the Civil War.[10]

The new program allowed students to choose any mix of the "parallel" courses and divided students into sections "...according to attainment or choice of studies," permitting the course work to be covered at different levels. Even those students, however, who were taking the traditional Classical Course were required to study more science than students at most other colleges.

Latin and Greek were omitted after the Freshman year in the Scientific Course, and the remaining three years were devoted about equally to science, mathematics (including differential and integral calculus), modern languages and social studies.

Natural philosophy had been a two-term course using Enfield's text until 1828. Starting that year, three terms of natural philosophy were required in both the Classical and Scientific courses. In the first two terms, students studied statics and dynamics using *An Elementary Treatise on Mechanics* by John Farrar of Harvard. Published in 1825, and based on works by Legendre and other French authors, this was the first of many translations and adaptations of French works in science and mathematics used in American colleges. The third term was devoted to lectures in electricity, magnetism and optics, based on a text by Biot, and the text for the required course in astronomy was also by Biot. The mathematics texts were mostly by French authors and by 1833, senior students in the Scientific Course were taking an additional term of advanced mechanics using Boucharlat's text.

Chemistry was decreased in the Scientific Course and two terms of natural history were added. The main text for natural history was John Ware's American edition of William Smellies' *The Philosophy of Natural History*.

When Joel Nott resigned as Professor of Chemistry in 1831, he was replaced by Chester Averill (Union, 1828). Shortly thereafter, there was a cholera epidemic in Schenectady, and Averill recommended to the Mayor that chloride of lime be added to the drinking water as a disinfecting agent. Although his advice was not taken, he was proved far ahead of his time, when that practice was first adopted in America in 1896. Despite his youthful promise, and early promotion to Professor of Botany and Chemistry, Averill died of tuberculosis in 1836.

## Student Opinions

Comments by students about new texts and new modes of instruction during this period of rapid change are contained in diaries and notebooks. Most notable were those written by Jonathan Pearson, who graduated in the Scientific Course in 1835; he remained at Union in various capacities for his entire life, leaving behind voluminous diaries covering all aspects of the College.

While an undergraduate, Pearson often commented about teachers and textbooks. One diary entry concerns the mechanics texts used in his junior and senior years:

> "Boucharlat's Mechanics is the most pleasing work I have studied this many a day on account of the clarity of his demonstrations in which a complete application of Analytical Geometry and Calculus is made, fully showing the superiority of his work over Farrar's."[11]

If Pearson had been exposed to the earlier Enfield's text, with its purely geometric "demonstrations," he certainly would have had some interesting comments.

Pearson became a tutor upon his graduation in 1835, then a faculty member, college treasurer and librarian. He served Union until his death in 1887. Although he is remembered today mostly for his diaries, equal recognition is deserved for a distinguished career in science education.

In a diary entry for June, 1838, Pearson reported that Dr. Nott had assigned him to teach chemistry, "...a study which I never before studied." This new undertaking was so successful that he became responsible for the bulk of teaching in chemistry and natural history at Union from 1839 until 1857, at which time he gave up his professorship in chemistry for one in agriculture and botany, but remained Professor of Natural History until 1873.

Meanwhile some other students, less interested in science than Pearson, resisted the large amount of science in Union's Classical Course, and in 1835 they requested the trustees to allow them literary electives in place of optics and astronomy. The trustees denied their request, stating that the "standard of attainments must yet be elevated" in these subjects.

A steady increase in mathematics requirements at Union and other colleges was also unpopular with many students. Leading colleges tried to require some calculus, but with little success; a Harvard official reported that "...one third of the class was making little or no progress" in the subject.[12]

Most of the basic changes in instruction in natural philosophy which led to the modern-day physics course were in place at Union by 1830. At this time Joseph Henry, who was then teaching at Albany Academy and becoming a renowned physicist, judged Union's philosophical apparatus comparable to that of Yale.[13]

An interesting description of a visit to view Union's natural philosophy apparatus is found in the diary of Asa Fitch, an RPI graduate who participated in a summer field trip by canal boat, led by RPI Professor Amos Eaton, up the Erie Canal. Their first stop was Schenectady, on July 2, 1830, where Fitch and several students knocked on Dr. Nott's door, with a letter from Professor Eaton. Nott was then president of both institutions.

Nott turned the visitors over to Professor Joslin, who conducted them around the apparatus room, with appropriate explanations and a few demonstrations. The following excerpts from Fitch's diary describe highlights of the visit (with paragraphing inserted to facilitate reading):

> "First the electric machines attracted our attention. They are very large, splendid and powerful and must (like most of the other articles) have cost immense sums....[There is] a very large one, with a flat glass plate for the cylinder—about 4 or five feet in diameter...They have, of course, a thousand amusing illustrations of electricity and magnetism...[We saw] a beautiful little brass steam engine....The cylinder on which the piston moves [is] about the size of a person's thumb. It is high pressure and whirls the wheel (Prof. J. said) with great rapidity...
>
> Then there were the Mechanic Powers—a large and..noble telescope...and there was a splendid Orrery...showing the relative size, distances, (but not in proportion to size), moons, etc. etc. All was moved...by a wheel in the bottom—the moon moving also...
>
> [There were] a kast [sic] head—to show phrenology, a French wired skeleton...different wheels for polishing and grinding down optics. Lastly, there was a musical clock, playing four tunes...Prof. J. wound it up—and then put a receiver over it & exhausted the air, by the air pump. The sound gradually lessened till it became a mere tinkling; and then increased as the air was let in again....
>
> On the whole they have a first rate suit of furniture, for complete demonstrations of Natural Philosophy and Chimistry [sic]..."[14]

Improved teaching methods and increasingly high standards slowly evolved at Union, utilizing the fine demonstration apparatus and better texts. Lectures continued to play a major role in the science courses, with students submitting their notebooks to the professor for inspection and grading. Supplementary recitation drills, since colonial times, had consisted mainly of students parroting what they had memorized from the text. This practice was gradually supplanted by a system in which students were required to answer questions and solve problems in recitation.

By the end of the 1830s it was common for Union seniors to give their commencement addresses on topics such as steam power and other scientific and technical subjects, instead of on common classical or civic concerns.[15]

## Back to Stone College

Once again overcrowding became a problem at Union. The Ramée Campus could not accommodate the rapid growth and the College was again bursting at the seams.

In 1831 Nott reacquired Stone College from the City of Schenectady to handle the overflow; in the interim it had been used as a City Hall and Jail. The Erie Canal had been constructed alongside the building early in the previous decade, and the Union Street bridge over the canal placed a major intersection at the College's door. It was now a busy downtown location.

Renamed West College, this building then housed freshmen, sophomores, the library, natural history cabinets and the Academy until 1854, when it was again sold to the City for use as a school.

## Professors Isaac Jackson and John Foster

Two new Professors of Natural Philosophy and Mathematics were appointed in the 1830s—Isaac Jackson (Union, 1826) and John Foster (Union, 1835). Both were creative teachers, multi-faceted and hard-working; each remained on the faculty for about 50 years, and had a profound influence upon Union, its students and curriculum.

Isaac Wilbur Jackson is best remembered at Union College today for his avocations: gardening and military training. He lavished years of work upon the extensive gardens later named in his honor which still beautify the campus. When still a student, Jackson became captain of the college drill team, a position he retained for many years. Known as "Captain Jack," he and his cadets participated in many commencements and college ceremonies; their uniforms and marching maneuvers were a colorful part of campus life.

**Isaac W. Jackson**

**Professor of Natural Philosophy & Mathematics, 1826–77**

*Jackson taught at Union for 51 years, beloved by students and fellow faculty. His creative mind sought order in many spheres: in addition to writing several textbooks, he trained the College drill team and developed "Jackson's Gardens" behind his residence in North College.*

Jackson's eminent teaching career deserves greater recognition. Born in 1804 of a Quaker family, he attended Albany Academy, from which he graduated with highest honors in mathematics and chemistry in 1824. Only two years later he graduated from Union, again with honors in math and chemistry.

He spent the rest of his life at Union College, serving as a tutor until 1831, then as Professor of Mathematics and Natural Philosophy. Although he was later designated the Nott Professor of Mathematics, he also taught physics until his death in 1877.

The Jackson family, including his mother, lived in a faculty residence at the north end of North College, adjoining his beloved gardens. He was responsible for overseeing the students who lived in adjacent sections of the dormitory, and managed them with humor, affection, and gentle discipline. His diary reflects concern for developing their minds and characters, as well as extending their intellectual attainments.

Jackson was a slight man, whose figure was enhanced by a handsome face, prominent brow and piercing eyes. Memorials attest to his generosity to friends and students, frank manner, social interests, and intellectual and professional attainments. Unfortunately, he suffered from ill health and depression all his life; his gardening activities were originally undertaken as a form of therapy, at President Nott's suggestion, and horticulture later became the delight and solace of his life.

Isaac Jackson was among the first of a new breed of college professors who devoted themselves entirely to the teaching of science and mathematics, without the distraction of other professional obligations, such as medicine or the ministry. Many years before the first graduate schools were established, they studied European texts and journals, mastering material long neglected in American colleges.

John Foster was one of Jackson's first students, graduating in 1835 in the same class as Jonathan Pearson. Only seven years younger than Jackson, he also spent his entire professional life at Union. After two years as a tutor, he was promoted to Assistant Professor, and then Professor of Natural Philosophy, replacing Benjamin Joslin who left Union in 1837 to accept an appointment at New York University.

After a long and productive teaching career, Foster was forced into retirement in 1885—a sad story told in the following chapter. Unfortunately, many files and letters relating to Foster are missing, so some important information about him is not available. However,

**John Foster**

**Professor of Natural Philosophy & Physics, 1836-85**

*Natural philosophy developed into physics during Foster's 49 year tenure. He was largely responsible for developing Union's outstanding collection of apparatus for demonstrations and individual experimentation, as well as developing courses which were precursors of engineering and later, electrical engineering.*

it is clear that he was a remarkable teacher with a keen interest in his subject and his students.

More interested in apparatus than was Jackson, Foster devoted a large amount of time and effort to demonstrations of physical phenomena. Whereas Jackson might today be considered a theoretical physicist, Foster was an early experimentalist.

Known familiarly as "Jack," he was reputed to be genial and kind-hearted and to have an excellent sense of humor. Unusually industrious himself, he demanded and received good work from his students, although a few considered him unduly demanding and overly critical. His large repertoire of demonstrations enlivened many lectures. These demonstrations were considered so valuable that on several occasions alumni responded to appeals by Foster to contribute funds for the purchase of more apparatus.

Foster also took great interest in civic affairs, and was often asked to present public lectures. His gift for sarcasm and biting wit were usually appreciated, but shocked his audience on at least one occasion. When he was invited to present a Fourth of July address in the early 1850s, to a large gathering of townspeople, his listeners anticipated an eloquent patriotic tribute. Instead, he devoted his address to criticism of the Schenectady public school system, denouncing, in scathing terms, a recent local practice of utilizing old meat markets for city schools:

> "...Tell it not in Troy, publish it not in the streets of Albany, that the people of Schenectady use their markets when they become too old for depositing in them the bones and flesh of animals, as places for depositing the bones of their bones and flesh of their flesh..."[16]

Shortly after Foster delivered this "thunderbolt," West College was again sold to the city. It became the central building in a newly established free school system, and also housed an "academical department" which replaced the Schenectady Academy. Union College paid the salary of the principal of this department and gave free tuition to its students who attended Union.[17]

Under Jackson and Foster's leadership, and that of other dedicated teachers, Union continued to attract large numbers of science students. By 1840, science courses at Union included instruction in botany, geology and mineralogy, anatomy and physiology, as well as astronomy, chemistry and natural philosophy.

Union ranked second in the number of graduates among thirteen northeastern colleges in 1842:

**GRADUATES OF NORTHEASTERN COLLEGES IN 1842:**[18]

103— Yale
93— Union
56— Dartmouth
54— Harvard
42— Princeton
36— Brown
34— Williams

30— Columbia
27— Amherst
23— Hamilton
21— City University of N.Y.
19— Rutgers
13— Middlebury

## The Development of Physics

In 1840, the college reorganized its expanding curriculum into eight departments; physics and chemistry were separated from mathematics and other sciences and grouped together under Physical Science.

Natural philosophy had developed into several specialized courses and the term "physics" was just coming into use, covering instruction in statics, dynamics, hydrostatics, pneumatics, electricity and magnetism and optics. Wave motion and sound were yet to become a significant part of the curriculum. Until this time it had been common for chemistry courses to include heat, light, electricity and galvanism under the heading of "imponderable agents." This practice was gradually discontinued as these topics were included in physics.

Although science students at Union covered more advanced topics in physics than those in the Classical Course, the latter were still required to study more physics than at most other colleges.

Classroom demonstrations became an increasingly important part of physics instruction, but the conditions under which the apparatus was stored were deplorable. John Foster complained to the trustees about the dampness in apparatus rooms in North Colonnade.

He also wrote a letter to alumni, appealing for support to upgrade the apparatus and succeeded in raising sufficient funds to enable purchase of a large amount of new equipment. In the early 1840s, Foster ordered apparatus from European instrument-makers, and from at least five dealers in this country, including Benjamin Pike of New York, George Dexter of Albany, and Phelps and Gurley of Troy. Among the items purchased were a battery, an electromagnet, a lift pump, a barometer, Nicholson's hydrostatic balance and a Daguerreotype Apparatus.

Dr. Benjamin Brandreth was one of those who responded generously to Foster's pleas for apparatus support. The Trustees' Minutes of July 21, 1846 recorded their deep appreciation to him for presenting to the College:

> "...a large and powerful air pump made by Benj. Pike of N. York for exhibition at a recent Fair of the American Institute and probably fully equal in finish and efficiency to any one which has been manufactured in the country."

These minutes also reported that:

> "...the same gentleman has also furnished the means for ordering from Paris a full set of apparatus belonging to the science of acoustics."

This order to Pixii totaled over $200, equivalent to $3,000 in 1994. It was a remarkable collection devoted to the newly emerging field of sound, including a pedal-operated bellows to demonstrate resonance in organ pipes (at $90, the most expensive item on the list), Chladni's apparatus to show standing waves in plates, Savart's Bell (another resonance demonstration) and an acoustical siren. Several of these instruments are still in the Physics Department collection and some have been used in demonstrations for 150 years.

Foster was advised on these purchases by Joseph Henry, then on the faculty at Princeton, who loaned him a Pixii catalogue. Henry had become a boyhood friend of Isaac Jackson (when they were fellow students at Albany Academy) and their close lifelong friendship later included John Foster.

On October 25, 1845, John Foster wrote a long letter to Henry describing his adoption of the "lecture" system for some of his courses, particularly a new course in electricity, magnetism and acoustics. He described his method as similar to that used by Henry and went on to describe it as follows:

> "Before commencing each lecture I examine them *viva voce* on the preceding and in addition require them to write out as fully as possible their notes and present them for examination once a week. I find it a laborious business to examine and correct some 80 or 90 books per week. Thus far the lecture system seems to work well—it evidently leads to more thought—creates a spirit of investigation and secures improvement in accuracy of expression. Still in such branches as mechanics where there is less of illustration by experiment and more of Algebra, I doubt the advantage of throwing away the textbook."[19]

**"Professor Foster Lecturing on Sound"**

*William Chambers, one of Professor Foster's students in 1847, pasted this amusing illustration into his class notebook.*

Among student notebooks in the College Archives is one by William Chambers, who studied sound and electricity and magnetism under Foster as a senior in 1847. His notes are sprinkled with humorous comments on the large number of demonstrations performed by Foster, who obviously was enjoying his new apparatus.

Astronomy was a required course for all students, and was taught by both Foster and Jackson. In the late 1840s, a small building was erected on campus to house a five-foot Dolland telescope. It probably was the same telescope later referred to by W.W. Edwards, who graduated in 1850. Edwards recalled:

"The College possessed a little three inch [diameter] telescope, which was kept in a little brick structure on the Campus between the North and South Colonnades. Some students in Astronomy, after some solicitation, persuaded Prof. Foster to give us a look through the telescope, which no one seemed to use."

After describing what a "great treat" it was to observe the moon, he remarked:

> "Practical Astronomy was not taught at the College of that time to any extent. What was then taught as Astronomy was rather Celestial Mechanics or Mathematical Astronomy."

Once again by the 1840s, textbooks had become outdated and deficient in coverage, so Jackson and Foster prepared their own lecture notes, drawing new information and new mathematical techniques from journals and foreign texts. All Union students took two terms of mechanics using Farrar's text, one term of optics using Brewster's text, and one term covering a variety of topics in electricity, magnetism and sound, using notes by Foster. As in Pearson's student days, a small group of students took advanced mechanics, using texts by Boucharlat and Poisson.

Comments from one student's journal describe reactions to courses from Jackson and Foster. The notes were written by Lemon Thomson, Class of 1850, in the fall of his senior year. After two terms of mechanics in his junior year, he was taking optics with Jackson ("Captain Jack"); electricity, magnetism and acoustics with Foster ("old Jack"); and elements of criticism with President Nott. Since these courses were required for all students, his experiences must have been fairly typical:

> "As soon as the King of day had penetrated the misty vapor which had enshrouded the Valley of the Mohawk, the Chapel bell was calling us forth from our rooms. Then we were hurried to recitation to try to explain old Jack's dry theories of musical vibration. There were several who bolted and some most extensive fizzling. We were bored for an hour and a quarter without our breakfast.

> ...The lesson in Optics hard. Capt. Jack scolded like an old woman; said students did not now study as they used to when he was in College....'Everyone who ever intended to do anything by study must have regular hours for study and must keep those hours sacred....He might lay up for himself a capital which in after life might be durable riches—or he might slide along through College and come out an intellectual pauper.'

> ...The weather being favorable we had a lecture and experiments on Electricity. We had a new lightning machine, it was a huge one—and it took a Paddy to turn it. The experiments were strange and interesting.

...At 11 attended recitation on Capt. Jack's visionary Wave Theory, this is supported by facts which are insupposable, and every step of the demonstration only makes it the more obscure. The Capt. makes this theory his idol and worships it on the same principle I suppose that a mother thinks the most of her most ugly deformed children.

...At the afternoon lesson we listened again to one of Jack Foster's interesting lectures on electricity—and they would truly be interesting if they did not cost so much labor to write them out—but this is truly Hurculean [sic] to write out a lecture every day—but to the candid thinker it must be a good discipline of mind besides affording useful knowledge."

Lemon Thomson afterwards became a Union Trustee, and in later years made generous contributions to the Library and to Foster's apparatus fund.

Lecture notebooks and performance in recitation were the primary bases for grades each student received in a course. This long-established pattern was common in all colleges and underwent little change until introduction of written examinations after the middle of the nineteenth century.

In 1848 Isaac Jackson published a text based upon extensive lecture notes; it was entitled *Elementary Treatise on Optics* and became so popular that it went through three editions. It replaced Bache's edition of Brewster's *Optics* as the standard text in American colleges and was also used in one British university. His *Elementary Treatise on Mechanics* was published in 1852 and went through four editions. He also published texts on trigonometry (1859) and conic sections (1836). The latter was extremely popular, going to eight editions.

### Savart's Bell

*This apparatus was developed for demonstrating acoustic resonance. Purchased in 1846 from Pixii, Paris, it is still in good condition in the 1990s.*

### Textbook Illustration of Savart's Bell

*An illustration from Ganot's* Elementary Treatise on Physics *shows a bell and tripod stand which are very similar to Union's apparatus.*

## Introduction of Engineering

**William Gillespie**

**First Professor of Engineering, 1845–68**

*Pioneering in establishing a new curriculum at Union, he integrated engineering with liberal arts.*

Although Union's new Scientific Course provided college students with some "practical" education, there was steadily increasing demand for more technically trained industrial leaders. After completion of the Erie Canal in 1825, other canals, roads and bridges were constructed throughout the state. The Mohawk and Hudson Railroad's "Dewitt Clinton" had pioneered steam-powered passenger service on its 1831 trip from Albany to Schenectady and the region had become a rail center. Eliphalet Nott had patented boilers which were providing clean and efficient steam power for Hudson River liners. These and other developments in industry and transportation led Nott to his last great curricular innovation—the Engineering Course.

Nott began this last of his major curricular experiments at the age of 72, with a background uniquely appropriate to the task. Whereas other college presidents lived mostly in a cloistered academic setting and stoutly resisted the introduction of "utilitarian" studies to their campuses, Nott had been personally involved in technology and industry for many years.

Incredible as it seems today, Nott also found time to serve as President of the Rensselaer Institute (later Rensselaer Polytechnic Institute) from 1829 to 1845. The Institute provided vocational training for students who did not wish to pursue a college program. In 1835 the Institute began offering an applied one-year course leading to a diploma in Civil Engineering, thus becoming the first civilian engineering school in the United States. When the Institute was reorganized in 1850, this was expanded to a three year course.

However, Nott foresaw a clear need for better-educated engineers with a stronger background in the liberal arts, science and mathematics. Farsighted as usual, he had begun to move Union in that direction in the early 1840s, reintroducing required courses in surveying for all science students and optional courses in topography and leveling. These courses were taught by John Foster.

A new program in Civil Engineering appeared in Union's Catalog for 1845, as an option for students in the Scientific Course. It occupied most of the junior and senior years and was the only engineering program before the Civil War to be successfully integrated into a liberal arts curriculum, leading to the B.A. Degree.

A new professor was needed to carry out this work, one with the necessary technical background but also with a thorough grounding in the humanities. William Gillespie fitted the role admirably and was appointed to head the new program in 1845.

Gillespie had graduated from Columbia at age eighteen and then spent ten years abroad traveling and studying at the Ecole des Ponts et Chaussees in Paris. He designed Union's unique engineering program, stressing a solid foundation in science and mathematics as well as the liberal arts. He also emphasized the importance of new technology by delivering a series of "popular lectures" on a variety of practical topics related to engineering, and all students in the College were required to attend.

In September 1844, Gillespie outlined his ideas for a new engineering curriculum in a long letter to Alonzo Potter, Union's Vice President. He asserted that the program would bring in more students, as well as strengthen scientific studies at the College, and he was correct on both counts.

Union's Civil Engineering program became a separate curriculum in 1855, offering its own degree, and became the country's second largest source of civilian engineers, after Rensselaer Polytechnic Institute, until the Civil War.

## Early Engineering Apparatus

Two historic instruments associated with William Gillespie and early engineering, a sector and a solar compass, have surfaced recently after being lost and forgotten for many years.

The sector, early precursor of the slide rule, was invented by Galileo as an aid to calculation. Union's sector is of brass, and was made by Butterfield of Paris about 1700, thus making it the second oldest instrument in Union's collection, after Gunter's quadrant.

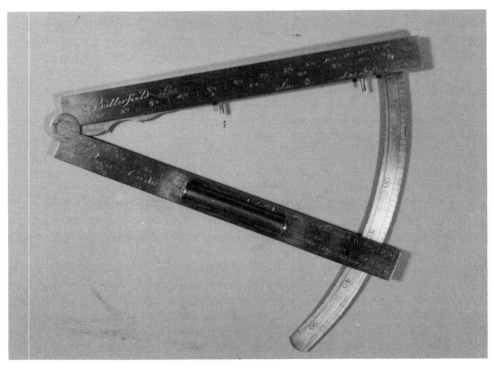

**Butterfield's Sector**

*An early aid to calculation, using the method of proportions, sectors were in general use throughout the 17th and 18th centuries. Union's sector was made in Paris about 1700.*

A mysterious package containing the sector arrived at Schaffer Library in the spring of 1990, with no return address. It was accompanied by an unsigned note which said:

> "Can't remember where I found this. The basement of some building there. It was never mine."

The instrument was turned over to Ellen Fladger, College Archivist, who asked me to help identify it. Despite heavy tarnish, careful polishing revealed a half-dozen scales, with labels in French. Medieval alchemist signs were deciphered for the metals gold through tin on a scale of relative weights for a given volume of each metal—a useful tool in many fields. It also displayed a small spirit level and a graduated arc for measurement of angles, with the maker's name, "Butterfield," on one of the hinged arms. (Michael Butterfield was an Englishman who settled in Paris in 1677 and was noted for the manufacture of fine scientific instruments.)

After consulting tomes of old scientific apparatus, I finally identified the instrument in a book which had once belonged to William Gillespie; his handwritten notes annotated many pages devoted to the use of the sector in the 1723 edition of Nicholas Bion's *The Construction and Principle Uses of Mathematical Instruments.* Gillespie had acquired much of his library and apparatus during a ten-year residence in Paris, and Union bought all of it, including the sector, from his widow after his death. We still do not know how long the sector was missing from the College, but are thankful it was returned.

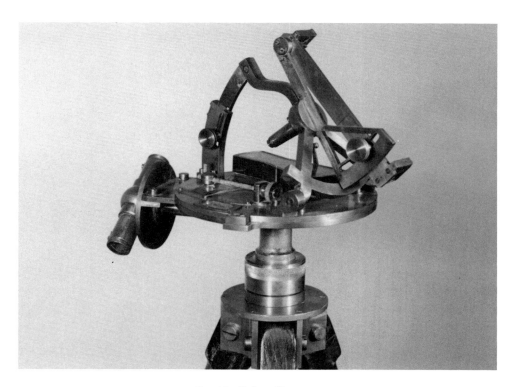

**Burt's Solar Compass**

*Union's model of this rare surveying instrument was made about 1840 by William J. Young of Philadelphia.*

The solar compass was patented in 1836 by Austin Burt, U.S. Deputy Surveyor, to overcome problems encountered in surveying areas with large magnetic disturbances. Burt exhibited his solar compass at the Great Exhibition in London's Crystal Palace in 1851. Union's model was made by William J. Young of Philadelphia about 1840, and is an outstanding example of the work of this early American craftsman. Union's solar compass was exhibited in the Union College Historical Exhibition of 1934.

## The Olivier Models

Another very important legacy from Gillespie is a set of 40 Olivier models, which illustrate the intersection of geometric surfaces by means of silken threads attached to brass frames mounted on a wooden box. Unlike earlier static models, the figures can be rotated about one or more axes, and thus represent a variety of geometric configurations useful in descriptive geometry.

They are esteemed for their elegance of design and construction, as well as their historical importance. In 1856 they were described in Jonathan Pearson's diary:

> "...really wonders of ingenuity and skill, the envy of all scientific men who see them."

The models were designed by Theodore Olivier, Professor of Descriptive Geometry at the Conservatoire Nationale des Arts et Metiers in Paris. Union's is the original set, constructed by Pixii in the 1830s in Paris, under the personal supervision of Professor Olivier. Gillespie had studied under Olivier, and after Olivier's death, Gillespie purchased the models from his widow in 1855.

In the 1960s, the models were restored to their original condition by the late Professor William Stone and are now on permanent display. Stone also wrote a monograph entitled *The Olivier Models,* which was published by the Friends of the Union College Library in 1969; it contains their detailed history and description.

**Olivier Model**

*Displaying the intersection of two cylinders of unequal radii, this is one of 40 wood and string models built for Professor Olivier to serve as teaching aids in descriptive geometry.*

## New Science Facilities and Developments

By the early 1850s, science and engineering departments at Union were again in need of more space and more modern facilities. Ever creative in financing, Nott provided the necessary funding for two new buildings, using income from the State Lottery and a variety of investments.

The first of the new science buildings was Philosophical Hall, located at the eastern end of North Colonnade. Completed in 1853, it was 50 by 80 feet long and two stories high. The Chemistry Department was housed on the first floor and the Physics Department on the second.

Drawings by William Gillespie show that the physics facilities included a 38 by 34-foot lecture room and an adjoining apparatus room of equal size. These occupied the entire front, or south, side of the building and provided a facility for physics demonstrations which must have been truly spectacular. The two rooms were connected by a fifty-foot railroad track which carried a 5 by 18-foot lecture table, allowing demonstrations to be set up and tested ahead of time and then quickly rolled into the lecture room. The rear section of that floor contained six smaller rooms, one of which was a workshop and one devoted to "Galvanism."

Soon thereafter Geological Hall was erected opposite Philosophical Hall, at the end of South Colonnade, to house geology and natural history over a chapel. This building is now known as Old Chapel.

The 1850s saw a steady increase in the amount of physics required in the Scientific Course, the Engineering Course and a newly-established program in Analytical Chemistry. By 1860 students in these courses had no time for classics and their diploma was printed in English, rather than in Latin.

Under provisions of a Trust Fund established by Nott in 1854, Isaac Jackson was named Nott Professor of Mathematics and John Foster became Nott Professor of Natural Philosophy. Each continued to teach both subjects but Foster managed the affairs of the Physics Department.

In the next few years, chemistry expanded rapidly at Union. Charles A. Joy (Union, 1844) had received a Ph.D. at Göttingen, and he was appointed the first Nott Professor of Chemistry in 1854. He left in 1857 for a professorship at Columbia University, after recruiting Charles F. Chandler as his successor.

Chandler was a dynamic young scholar, also with a Ph.D. from Göttingen. Under his leadership Union opened the first Analytical Chemistry program in America with a laboratory equipped specifically for undergraduate use. In 1861 the first student chemical society was founded at Union, a precursor to the American Chemical Society. After seven years, Chandler left Union to join Joy at Columbia in 1864.

Maurice Perkins, a graduate of Harvard, was the next Professor of Chemistry, and he remained at Union until his death in 1901.

## Seeds of Trouble

Although new facilities were provided on campus for science, the library was overlooked. Displaced from West College, the collection of books was stored away in boxes for several years. Finally some space was found for them in a small room above the chapel; adequate library facilities were a long-deferred need.

By 1854, when a thousand alumni assembled in Schenectady to celebrate the semi-centennial of his presidency, 81 year old Eliphalet Nott was crippled by arthritis and declining rapidly in his ability to lead the college. However, he would not let go the reins and his vice president and presumed successor, Laurens P. Hickok, was unable to assume any real authority in the operation of the College.

In 1859 Nott suffered a paralytic stroke from which he never fully recovered, but he still retained the presidential title. From then until his death in 1866, the college began a downward spiral. The decline would not reverse until near the end of the century.

**Isaac Jackson, his Horse (Cosine) and his Dog (Beauty) were a familiar sight on campus for many years.**

# CHAPTER FOUR

## YEARS OF DISCORD: 1861–1885

The Civil War ushered in a long period of difficulties for Union College, beset by campus problems as well as the national trauma.

Enrollment declined by fifty percent as students and alumni joined both sides in the conflict. Initially, Union retained an advantage over other colleges, because of the popularity of its Science Course; in 1863, it had 205 students, while Columbia had only 186. Renewed demands for soldiers, however, took an increasingly heavy toll.

Isaac Jackson's oldest son, William A. Jackson (Union, 1858), organized and drilled a group of young men from the Albany-Schenectady area, called Zouaves. They became Company A of the 18th New York Volunteers, with Jackson as their commanding officer. A group of Union College Zouaves was organized by Elias Peissner, popular young Professor of Modern Languages, who also led his company into battle. Both Peissner and young Jackson were killed in the war, along with many other Union men.

After Eliphalet Nott's death in 1866, the College entered a troubled era brought on by devastating effects of the war, lack of sustained leadership, and serious financial problems. College resources and Nott's personal finances had become terribly entangled; successors could not continue his legerdemain.

Enrollment failed to increase after the war. In the brief span of twelve years, from 1860 to 1872, it fell from 437 to 89.[1]

### Leadership Problems

Nott proved to be irreplaceable. The trustees had never developed any leadership or responsibility for management of the College as Nott had pre-empted that role. Afterwards they seemed to believe that nothing more was required of them than appointment of a new president.

Laurens P. Hickok (Union, 1820), who had been Vice President of Union since 1852, was the logical candidate for the job, but was opposed by many faculty, alumni and trustees. Hickok had served as Professor of Mental and Moral Philosophy, and made important contributions to ethics, epistemology and metaphysics. In 1858 he had published a book entitled *Rational Cosmology,* which was criticized by Isaac Jackson because of scientific errors. Jackson became one of the leaders of the anti-Hickok faction, which also included Foster and Gillespie.

Despite widespread and bitter opposition, Hickok was appointed to succeed Nott in 1866. He was described by Jonathan Pearson as:

> "A good, honest man with many qualifications for his office...he
> has a strange want of tact...no power to conciliate..."

After serving for two difficult years as president, Hickok retired in 1868.

Charles A. Aiken was the next president, and he succeeded no better than his predecessor, also remaining for only two years. By the 1870s, college faculties were starting to see themselves as independent professionals, and refused to be treated like mere employees whose first loyalty was to their president, rather than to their profession. They did not hesitate to disagree with administrators, trustees, or with each other. Factions and enmities developed which divided the campus for many years.

The next president, appointed in 1872, was Eliphalet Nott Potter, son of Alonzo Potter and grandson of Eliphalet Nott. Potter's entire twelve years in office were also marked by controversy, as he attempted unsuccessfully to continue the autocratic one-man rule so effectively conducted by his grandfather.

During his term, four separate institutions were brought together to form Union University in 1873. They were Union College, Albany Law School, Albany Medical School and Dudley Observatory.

The Observatory had been founded in 1852 as part of a projected "National" university to be established in Albany. Since other components of that university never materialized, the Observatory's first twenty years had been very difficult.

At the time Dudley became affiliated with Union University, its director was George Washington Hough (Union, 1856), a distinguished astronomer. Hough carried out an active program of astronomical observations during his fourteen years at Dudley, and also developed a number of instruments for meteorological studies, including the first successful recording barometer.

There does not seem to have been much interaction between the College and the Observatory as a result of this affiliation, although the College's catalog for 1880 contained the following statement:

> "Astronomical and Meteorological observations are regularly made at the Dudley Observatory. The apparatus is large and rare. Instruction is given to special students."

**George Washington Hough**
**Director of Dudley Observatory, 1860–1874**

*Hough was an alumnus of Union College (1856).*
*After leaving Dudley, he became Professor of*
*Astronomy at Northwestern University.*

Faculty protests against President Potter finally culminated in a "trial" before the Board of Trustees and his subsequent dismissal in 1884.

E. N. Potter is best remembered today for his success in raising funds for apparatus and buildings, including the long-delayed completion of the Nott Memorial, with his brother, Edward T. Potter, as architect.

## The Nott Memorial Building

Ramée's original plan for the campus had included a central rotunda with a high dome at the focal point. When the campus was first built, all available resources were expended upon construction of North and South Colleges, but Nott had always intended to complete the plan when he could raise sufficient funds.

In the late 1850s, he had founded an alumni organization and started raising money for construction of the central building, which he then called "Graduates Hall." The ground floor was planned as a chapel and convocation area, with an alumni hall above. In 1858, with only $4,000 in pledges, Nott organized a cornerstone-laying ceremony to celebrate the start of this long-delayed construction.

Edward Tuckerman Potter (Union, 1853), already established as a talented and innovative architect, was chosen to prepare the building plans. He adapted Ramée's original plans for a round building into a 16-sided polygon, and incorporated Ramée's arcade features into the new design.

59

# EXERCISES

## PERFORMED AT THE BURIAL OF

# MECHANICS,

## BY THE JUNIOR CLASS OF UNION COLLEGE,

### JULY 13, 1863.

———————

In accordance with a custom recently inaugurated in this College, the Class of '64 assembled to pay the last *earthly* tribute to *Mechanics*, with whom their intimate acquaintance has now ceased *forever*.

On Monday evening, July 13th, at eleven o'clock, the funeral procession was formed under the supervision of Marshal James L. Seward, assisted by Aids M. M. Skiff and B. G. Smythe.

The following was the

### ORDER OF PROCESSION.

Brass Band, 13 pieces, led by Geo. Mayer.

High Priest.

Orator.                                    Poet.

Torch Bearers.        Execrator.        Torch Bearers.

Hearse.

Pall Bearers.

# C. O. W.

Class of '64 with torches, banner, and transparencies.

**A Program from the annual ritual of "The Burial of Mechanics"**

*The ceremonial burying of textbooks was a popular practice for many years.*

In 1859 a stone foundation was erected, with walls rising several feet above ground, to provide for a roomy and well-lighted basement. Work proceeded slowly through the spring and summer, but ground to a halt in the fall, when available funds were all expended. A hiatus ensued for the next 13 years, due to Nott's illness and death, the Civil War, and a widespread depression, all of which interfered with substantial fund-raising.

Work resumed in 1873 and was finally completed in 1877. A number of design changes reflected Potter's more mature conceptions, and advances in architectural technology. Many advanced features were incorporated into the design, such as cast-iron columns supporting two galleries, the dome and the clerestory.

A remarkable ornament set off the dome with 709 "illuminators," pieces of colored glass, each about two inches in diameter. In the southern half of the dome, illuminators were arranged in a Newtonian spectrum with red at the base and violet at the crown. In the northern half, the illuminators were red, rose, lavender, blue and violet. It has been suggested that these two sets symbolized the "scientific" and the "sacred," and that their combination or union in the dome represented one of the early goals of Union College.

For further information on such features and their interpretation, see a fascinating unpublished manuscript by Carl George and Robert Uzzo entitled "The Nott Memorial—Symbolic Elements in the Architecture of Edward Tuckerman Potter."

## The Four Year Science Curriculum

Through all the administrative chaos and faculty dissension, Jackson and Foster continued development and improvement of the science curriculum.

In 1865 the Science Course was "remodelled" and extended from three years to four, bringing onto the campus the freshmen science students who had been relegated to the Union School in the former West College since 1856.

The catalog announced that scientific students were now to spend all four years on campus, taking a more rigorous course as well as intensive study of modern languages extending through the junior year. Their diplomas, which had been in English since 1852, would now be in French; this probably indicated that Francophile William Gillespie played a role in the planning of the new program. According to the catalog, the new science course would now be:

**Newton's Seven Mirror Apparatus**

*Union's device for separating and recombining the colors of the spectrum was made by Jules Duboscq of Paris about 1870.*

"...a four year's course, intended to be fully equal in amount of study and in disciplinary value, to the classical course; with which it now runs *pari passu*. This change has been made in compliance with increasing demands for a full course especially calculated to fit young men for the higher walks of a scientific, commercial, political or diplomatic career."

Few additions had been made to the physics apparatus during the Civil War, but in 1867 John Foster traveled to Europe and purchased a number of items. In 1870 he again appealed to alumni to provide funds for further acquisition of new equipment, and he collected approximately $5,000, equivalent to at least $50,000 today. In a report to President Potter, dated May 22, 1872, Foster included a long list of apparatus added to the department during the preceding two years, stating that these additions had been furnished "...chiefly from money generously contributed by several graduates of the college." He closed his report with the statement that an additional $5,000 would place the department in a very satisfactory condition as regards the means of instruction.

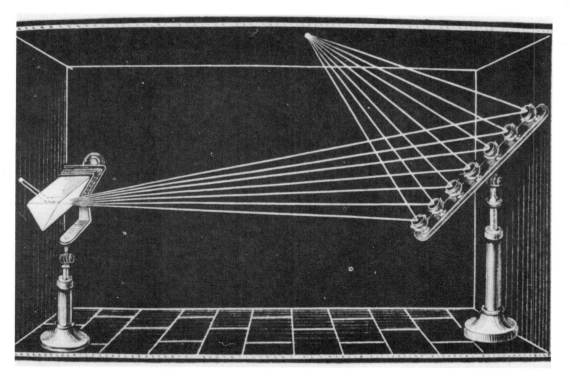

**Text Illustration of the Seven Mirror Demonstration**
*The light source was sunlight, emerging from a small hole in the wall.*
(*From Ganot's* Elementary Treatise on Physics)

Union had been a leader in the introduction of student laboratories in chemistry in the 1850s, and student laboratories in physics (distinct from demonstrations by the professor) probably began about 1870—earlier than has been generally recognized until now.

Individual laboratory work in physics had begun some years earlier in Europe, and was first adopted in the United States in 1869 at the newly-founded Massachusetts Institute of Technology. Union was one of the first colleges to follow M.I.T.'s example. The catalog for 1869–70 listed two terms of laboratory exercises in the senior year of the Scientific Course, without specifying whether they were in chemistry or physics. However, there is strong evidence that some of Foster's advanced students were doing independent laboratory work at that early date.

Some of Foster's new apparatus clearly was designed for use in a laboratory context rather than for classroom demonstrations. These included instruments requiring delicate adjustment and individual recording of data, such as spectrometers, polarimeters, refractometers and a wide variety of optical and electrical equipment. Further evidence of a laboratory context is contained in the 1872 report from

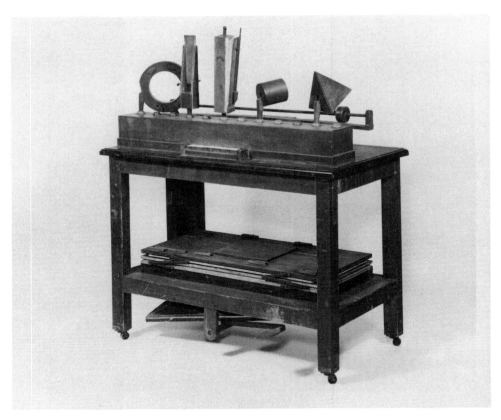

**Foot-Operated Bellows for Demonstrating Resonance
in Organ Pipes**

*This apparatus has a keyboard and space for twelve pipes. Union purchased it in
1872 from Rudolph Koenig of Paris, and donated it, in the 1960s, to the Smithsonian
Institution, along with several other valuable items.*

Foster to the president, stating that the apparatus room had been
provided with a variety of supports, batteries, gas tanks, etc., and
three large and handsomely finished walnut tables "...to facilitate the
use of the apparatus in experiments."

By 1875, physical laboratory exercises were listed separately in the
catalog as a senior option. Professor Isaiah Price (Union, 1872) may
have influenced Foster in this direction, as he spent the academic
year 1873–74 abroad, taking advanced studies. He undoubtedly
returned with reports of excellent student laboratories in several
European universities.

Foster was honored in 1873 by Columbia University, which conferred upon him an honorary degree of Doctor of Laws. Jackson had
been similarly honored earlier by Hobart College.

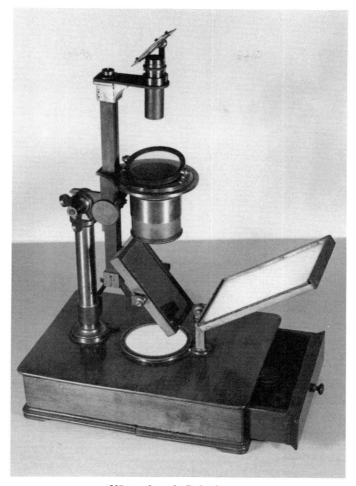

**Nörrenberg's Polariscope**

*One of several instruments acquired by Professor Foster to demonstrate polarization of light.*

During the winter of 1873, Foster suffered a collapse described as "severe nervous prostration" which he later attributed to a combination of overwork and lack of heat in the apparatus rooms.[2] During his illness, his classes were taught by Sidney Norton (Union, 1856).

Foster spent the following year recovering and traveling to Europe to procure more apparatus, mostly from Paris and London. He had been authorized to spend an additional $6,000 on laboratory equipment; after these purchases, Union's collection of apparatus was again recognized as one of the best in the country.[3] Foster devoted the fall of 1875 to unpacking and organizing the new equipment and giving many demonstration-lectures to visitors and groups from the surrounding area.

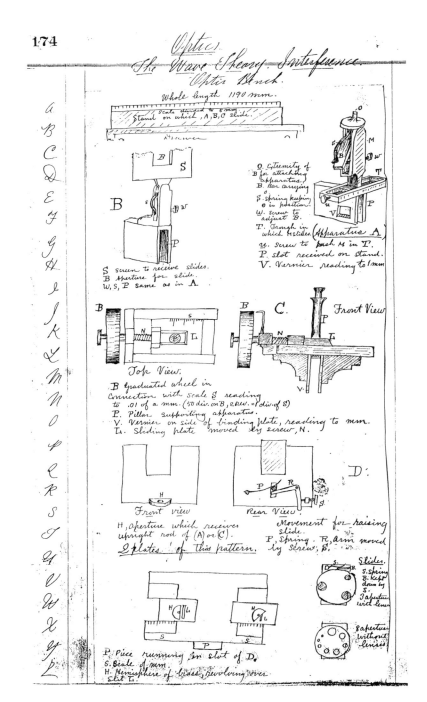

**Sample Page from "Union College Physical Laboratory"**

College archives from this period contain records of his purchases, and—more important—a very impressive two-volume Inventory of Apparatus entitled "Union College Physical Laboratory," which includes over 700 items. The inventory is undated but its contents suggest that it was prepared about 1875 under Foster's supervision. It is probable that two other staff members assisted in its production, Isaiah B. Price, Adjunct Professor of Physics and Sidney A. Norton, who served for a short period as Lecturer in Experimental Physics.

The Inventory contains detailed descriptions of many of the important pieces of equipment, along with manufacturers' names, catalog numbers and some prices. In addition, there are illustrations from catalogs and many hand-drawn sketches showing construction and use of the apparatus. This work is invaluable in helping to identify many mysterious and elegant old pieces which have adorned the physics stockrooms for many years, as well as providing a picture of the techniques by which important physical principles were presented to students more than a century ago.

John Foster published a textbook in 1877 which he had been using in syllabus form for many years. Entitled *An Elementary Lecture on Electricity, Magnetism, Galvanism, Electromagnetism and Acoustics,*[4] it attracted much favorable attention and was chosen for display at the International Electrical Exhibition in 1884.

## Jackson's Tributes and Death

Commencement weekend of 1876 included a celebration of Isaac Jackson's fifty years on the faculty of Union College. There were dinners and speeches honoring him for a lifetime of dedication to the advancement of Union and its students. A 32-page account of the proceedings, published by the College, reports that alumni presented him with a massive silver bowl at the end of the ceremonies.

An unknown member of the class of 1878, reminiscing many years later, reflected on the mood of the time:

> "Union then stood at the parting of the ways between the old and the new. The old professors were still there. 'Captain Jack,' most beloved of teachers, was still holding classes, although it must be confessed that few took them seriously, since the dear old man was too kindly to 'stick' any of us. All loved him and all took advantage of him. And yet much might have been learned of 'Captain Jack,' for he retained even then, after 50 years of service, the vigor of mind and clearness of expression that had made him a splendid teacher..."[5]

67

Isaac Jackson died shortly thereafter, on July 28, 1877, at the age of 73.

## The Clarke Survey

An extensive survey on the teaching of chemistry and physics in the United States was conducted in 1878 by Frank W. Clarke, Professor of Chemistry and Physics at the University of Cincinnati. Published by the U.S. Bureau of Education in 1881, it contained information on courses, textbooks and laboratories in more than 350 colleges and universities.

Union was one of only 34 institutions which reported any laboratory facilities for students in physics, although 138 colleges reported student laboratories in chemistry. It was common at Union and elsewhere that only advanced students were allowed in the laboratory. The report also stated that Union had laboratory desks for ten physics students at a time, an unusually large amount of space for a college of Union's size. Cornell, which had a much larger enrollment, reported that only ten or twelve of its students elected laboratory work in physics each year.

Professor Clarke's detailed analysis of the offerings in chemistry and physics revealed the sad state of instruction in those fields at most institutions. His conclusions are informative:

> "In the larger number of colleges...the studies in question occupy much the same position which they do in the ordinary high schools and academies....Many high schools are actually doing more and better work with these sciences than is done in a very considerable number of colleges bearing good reputations....The present subordinate position of scientific studies is undoubtedly due to the continuation of the old-fashioned plan of a fixed curriculum...with its emphasis on the classics....The colleges which are doing the best work in chemistry and physics are those which have adopted the elective system."[6]

Both an elective system and a scientific alternative to the classical curriculum had been a part of Union's program since 1828.

Clarke tried to classify institutions with regard to their offerings in chemistry and physics, based on the number of courses taught, the level of the textbooks used and the amount of laboratory work involved. He had difficulty with the physics classifications, however, pointing out that:

> "The returns sent in concerning this science exhibited curious variations, caused by diversity of opinion as to what should be included in it."[7]

By 1878 technical institutes and graduate schools had proliferated but, among the 350 institutions surveyed, fewer than ten required as much physics for graduation as Union; at this time, Classical undergraduates had to take one and a third years, whereas Science students took two years, plus senior electives in physics.

Professor Clarke found that the most popular text for the general physics course in the highly-rated institutions was Ganot's *Elementary Treatise on Physics,* translated from the French and edited by Atkinson. This remarkable work, which had gone through at least ten editions by 1878, contained 1,000 pages; 1,000 woodcuts were included, illustrating every type of apparatus known at that time. Union's Schaffer Library now has four editions of this text.

At that time Union College was using Jackson's *Elementary Treatise on Mechanics,* and his *Elementary Treatise on Optics,* as well as Foster's *An Elementary Lecture on Electricity, Magnetism, Galvanism, Electromagnetism and Acoustics.*

Two major differences stand out when these books are compared with the much more popular text by Ganot, which surveyed all of physics in one volume. First, the Union texts had been in use for over 25 years. Although they had been revised several times, their formats were from an earlier era; the illustrations were small line drawings on foldout plates in the back of the book, whereas Ganot placed his in the body of the text. Secondly, the mathematical level of Jackson and Foster's books was generally higher than that of Ganot, who used no calculus and little trigonometry. It would appear that, though less encyclopedic, Union's physics courses were more analytical than many of those at comparable colleges.

## John Foster's Last Years

After the death of Isaac Jackson in 1877, Professor Foster gradually became isolated from the faculty mainsteam. He attempted to remain uninvolved amid increasingly bitter disputes between faculty and administration. Immersing himself in work, he took on additional teaching duties and continued to raise funds for new physics apparatus.

In June 1878, he suffered a slight stroke which affected the nerves on the left side of his face and his left eye. Apparently, it was not debilitating, since he continued teaching and traveling on College business. In the spring of 1882 he addressed the Union College Alumni Association of the Northwest. On that occasion he was presented with $1,000 "...to purchase an electric machine for

Union College." The same issue of *The Concordiensis*[8] which reported this gift included the following item:

> "Fitzgerald and Landreth have been putting up an engine and boiler in the room adjoining the photographic room, for the purpose of running Prof. Foster's dynamo-electric machine to furnish electricity for his class experiments, thus obviating the necessity of filling and emptying batteries every day."

In 1882 he received $200 from alumni to purchase a large collection of Crookes tubes for the study of electric discharges in gases, and a few years later these tubes proved to be extremely useful for demonstrating x-ray phenomena. By the following spring Foster was teaching seniors two labs weekly, in addition to lectures. In 1883 he reported that the physics apparatus was valued at $15,000 and that the dust from the stoves used to heat the rooms was "exceedingly injurious." Once again he tried unsuccessfully to have the College install steam heat in the labs.

Foster continued to be innovative in his teaching, with new courses in practical electricity and electrical testing and he indicated in correspondence that he was considering "the ultimate possibility of full instruction in Electrical Engineering."[9]

Other forces were at work within the College, however, which were opposed to and now actively intriguing against him, because he had refused to support a faculty movement attempting to impeach President Potter (although he did support constructive actions to persuade Potter to amend unpopular decrees). Some colleagues misrepresented his position, exaggerated the extent of his health problems, and accused him of decrepitude. They threatened to retaliate against him.

Finally, in 1884, Eliphalet Nott Potter was removed from office and Judson Stuart Landon, a trustee favored by faculty radicals, became acting president. He immediately reorganized the Physical Science Department without consulting Foster and gave several of Foster's courses to Winfield Chaplin, Adjunct Professor of Physics.

On February 6, 1885, after nearly 50 years on the faculty, John Foster was dismissed without any prior notice. He was 73 years old. Chaplin took over the department but remained at Union only until the end of that year.

At first Foster was promised his full salary, in conjunction with a reduced teaching load, but actually was then fully retired at half-salary. (There were no provisions for retirement in those days.) Alumni were shocked by his dismissal and 500 signed a petition requesting that it be revoked.

Professor John Foster with Sigma Phi Brothers, 1882

His version of the events leading up to his sudden, unexpected retirement are contained in *An Open Letter to the Alumni of Union College,* in which he refuted the charges of infirmity and incompetence used to justify his dismissal. Foster wrote this defense shortly after his dismissal, but delayed publication until 1894, at the close of President Webster's administration, in order to avoid impeding "...efforts in progress to improve the financial condition of the College."

The precise facts in the matter probably never will be known, but from this writer's perspective, a hundred years after the event, Foster's *Open Letter* is very convincing.

Since subsequent histories of Union have all avoided discussion of this dark period in the life of the College, it is likely that very few people have read John Foster's seventy-page defense. However, at least one outside historian believed Foster's version of the story and published a summary account:

"At Union, when John Foster refused to join the faculty movement against President Potter, he reported that a young professor informed him that '...If I did *not* join them, I would certainly be destroyed.' As soon as President Potter left, the threat was carried out. A professor who had once been a student of Foster wrote to the Trustees that Foster had neglected his teaching duties and was teaching outmoded interepretations of physics. A secret letter was circulated among the Union faculty seeking to prove that Foster had at first indicated his willingness to sign the faculty petition against Potter, then had betrayed his promise....Foster was fired after 47 years of teaching, without a semblence of the inquiry he had requested to determine whether, in fact, he had been neglecting his duties..."[10]

After his forced retirement, Foster and his wife still lived in their campus home, just inside the Nott Street gate, across from the avenue which was named in his honor. Built in 1814 as a dining hall, his house had been known as North College Boarding House until 1860, when it was refitted as his residence.

On March 17, 1895, on a cold winter night eleven years after his retirement, Foster accidentally set his home on fire. He dropped an oil lamp while going upstairs, and the resulting conflagration destroyed most of the house and its contents.

Although President Raymond and the trustees conducted a fund drive to rebuild the home, work was not completed before Foster's death on October 21, 1897, at the age of 86. Fero House now occupies that site, and fire-damaged remnants of the 1814 structure are still visible.

John Foster deserves more recognition than he has received as the moving spirit behind Union's innovative courses in Natural Philosophy, Physics, and Civil and Electrical Engineering.

**A Centennial View of the Union College Campus, 1895**

*From The Centennial Souvenir. At that time, the physical sciences were housed in Philosophical Hall and North Colonnade.*

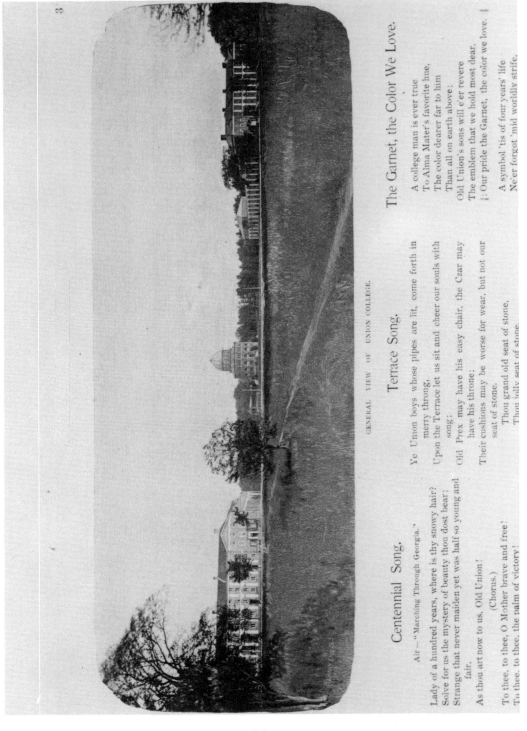

GENERAL VIEW OF UNION COLLEGE.

## Centennial Song.

Air—"Marching Through Georgia."

Lady of a hundred years, where is thy snowy hair?
Solve for us the mystery of beauty thou dost bear;
Strange that never maiden yet was half so young and
    fair,
    As thou art now to us, Old Union!

(Chorus.)

To thee, to thee, O Mother brave and free!
To thee, to thee, the palm of victory!

## Terrace Song.

Ye Union boys whose pipes are lit, come forth in
    merry throng,
Upon the Terrace let us sit and cheer our souls with
    song;
Old Prex may have his easy chair, the Czar may
    have his throne;
Their cushions may be worse for wear, but not our
    seat of stone.
    Thou grand old seat of stone,
    Thou jolly seat of stone.

## The Garnet, the Color We Love.

A college man is ever true
To Alma Mater's favorite hue,
The color dearer far to him
Than all on earth above;
Old Union's sons will e'er revere
The emblem that we hold most dear,
[ Our pride the Garnet, the color we love. |

A symbol 'tis of four years' life
Ne'er forgot 'mid worldly strife,

# CHAPTER FIVE

## TRANSITION TO A NEW CENTURY: 1885–1900

Union College's leadership crisis was finally resolved in 1888, when the trustees were spurred to action by massive student protests. The College had been drifting for four long years, while Judge Landon served as acting president, and morale was very low among faculty and students. Matters finally came to a head when most of the student body paraded through downtown Schenectady, calling attention to the College's plight. They adopted a formal resolution to "...withdraw all loyalty and allegiance to the College and transfer our interests elsewhere" unless a new president were appointed.[1]

**Thomas W. Wright**
**Professor of Physics, 1885–1905**

Shortly thereafter, Harrison Webster became president. He was a former Professor of Natural History who had been fired during the Potter controversy, and his return signaled an end to ruinous factional dissension. From that time on, major elements worked together for the good of the College.

The retirement of John Foster in 1885 ended an era in physics and mathematics teaching at Union—a remarkable period of more than half-a-century in which the team of Jackson and Foster had deepened and expanded these subjects.

Thomas W. Wright was appointed to fill John Foster's position in 1885 and he remained at Union until his retirement in 1905. Born in Scotland and raised in Canada, he was a graduate of the University of Toronto with an additional degree in Civil Engineering from the Sheffield Scientific School at Yale. He had worked for the U.S. Lake Survey for ten years, and had published *A Treatise on the Adjustment of Observations with Application to Geodetic Work and Other Methods of Precision*, a reference book intended for use in the field, although it was adopted as a text at Columbia.

Professor Wright immediately began a major rearrangement of courses and textbooks, a difficult task made more onerous by critical comments from Emeritus Professor Foster.

This was a time when the elective system was again taking precedence over required curricula, allowing more flexibility for individual interests and specialization. After the second term of the junior year, only half of the course work was prescribed, so students could choose from an expanded number of electives.

Two new advanced physics courses were offered as electives for seniors in Union's catalog for 1887. One was mathematical physics which included the method of least squares with application to physical problems, and probably used Wright's text. The second new course was physical laboratory, with special attention given to electricity, for which Foster had acquired a large collection of modern apparatus.

The 1888 catalog indicates that all students were still required to take astronomy in their senior year and three terms of physics in their junior year, but courses for science and engineering students were at a higher mathematical level than those designed for students in the Classical Course. Students in the Scientific Course were also required to take five terms of French and four of German.

Physics texts listed in the catalog emphasized electricity, with titles which later would be classified as Electrical Engineering, such as Ayrton's *Practical Electricity*, Thompson's *Dynamo-Electric Machinery* and Kempe's *Electrical Testing*.

A special Honors Course in Physics was introduced in 1889, starting in the third term of the junior year and extending through the senior year. It consisted of four terms of experimental work in the laboratory and three terms of mathematical physics.

In 1890, the Bachelor of Science degree was instituted for graduates of the Scientific Course and a Bachelor of Philosophy Degree (Ph.B.), sometimes called the "Latin Scientific Degree," was introduced for those who desired a combination of the Classical and the Scientific programs.

Thomas Wright published *Elements of Mechanics* in 1891, a well-organized text with many excellent illustrations and problems which made free use of calculus.

In 1894 Howard Opdyke was appointed as a second physics teacher. A graduate of Williams College, he had studied an additional year at the Columbia School of Mines. He was to remain at Union until his death in 1928.

The mid-nineties brought more curricular changes, reflecting trends common to colleges across the country. Physics lost place as an essential college study; Yale and many other excellent colleges dropped it as a requirement for graduation. In 1897 Union reduced required physics courses for Classical students from three to two, and astronomy became an elective.

At the same time, however, physics was increased for science and engineering students; four courses were required, including three terms of laboratory. This was another major change, as physics laboratory had previously been an elective. Astronomy was still required for these students.

When President Webster retired in 1894, Union was on the upswing but was deeply in debt. President A. Van Vranken Raymond took office hoping that the approaching College Centennial Celebrations would bring forth generous gifts to solve that problem. Unfortunately there was a severe economic depression at that time, and no major gifts were forthcoming.

Union's financial situation was desperate and the College almost on the verge of closing, when powerful citizens in Albany proposed that a large amount of money and land for a new campus could be made available if Union would relocate there.[2]

Then, almost miraculously, Eliphalet Nott came to the rescue, as if he had risen from the grave! A Nott Trust Deed, enacted some 40 years earlier, had conveyed to the College a large tract of land in what is now Long Island City. Through the years, taxes on this property had been a heavy strain on College finances, and in 1897 the trustees finally decided to sell it. To everyone's amazement, the property was worth much more than anticipated, and funds from its sale paid off all the College's debts and left $300,000 to begin an endowment.

**Rooms in Physical Laboratory, circa 1900**

*(above) Physics Demonstration Classroom*
*(below) Apparatus Room and Laboratory*

It was still necessary, however, to make painful cuts in order to put the College on a balanced budget. Faculty were dismissed, department heads took salary cuts and all operating expenses were reduced. Then several large bequests were received, and the College was finally on a sound financial footing.[3]

Soon thereafter a long-delayed building program was underway. It included complete remodeling of North and South Colleges into modern dormitories, renovation of the Nott Memorial Building into a first-rate library, refurbishing and re-equiping of chemical and physical laboratories, and the erection of two new buildings, Silliman Hall to house student activities and an electrical laboratory.

In 1895 a separate program in electrical engineering was added to the curriculum. This resulted from a conference between the College and the General Electric Company, newly-formed from the earlier Edison Machine Works in Schenectady. In 1901 Charles Proteus Steinmetz would be appointed Professor of Electrical Engineering, on released time from General Electric, and the program would expand rapidly.

The growth of General Electric and the American Locomotive Company led to explosive growth in the city of Schenectady. Between 1890 and 1907, the population surged from 19,000 to 70,000.

Similar growth would take place at Union. In 1891 there were only 12 graduates, five of whom studied engineering. In 1905 there were 29 graduates of whom 18 were engineers, and in 1909 there were 92 graduates, more than half of whom were engineers.

From the nadir of its centennial period, Union was thus spectacularly revitalized. Shortly after its second century began, Union was poised to prosper in a new age, with a scientific curriculum at its core which still reflected Eliphalet Nott's ideals of education.

# APPENDIX A

## UNION'S FIRST CURRICULUM

**From the *Laws of Union College*, printed in December, 1795.**

The sequential classes were called The Class of Languages (first year), Class of Belles-lettres (second year), Mathematical Class (third year) and Philosophical Class (fourth year). French could be substituted for Greek in all four years.

## CHAP. IX.

### *Education.*

#### SECTION 1.

THE Students in the Clafs of Languages, fhall be acquainted with Virgil, Cicero's Orations, Greek Teftament, Lucian, Roman Antiquities, Arithmetic, and Englifh Grammar; or, inftead of the Greek, with Gil Blas, in French, before they can be admitted into the Clafs of Belles-lettres.

#### SECTION 2.

Before admiffion into the Mathematical Clafs, the Students fhall be acquainted with Geography, and the ufe of the Globes, the Roman Hiftory, the Hiftory of America, and the American Revolution, Chronology, three books of Xenophon, Horace's Odes and Satires, and Criticifm; or, inftead of the Greek, with a Hiftory of the French Revolution in French.

#### SECTION 3.

In order to admiffion into the Philofophical Clafs, the Students fhall be acquainted with Vulgar and Decimal Arithmetic, and the Extraction of the Roots, Geometry, Algebra, Trigonometry, Navigation, Menfuration, Xenophon continued, and Homer; or, inftead of the Greek, with Bofuet's Univerfal Hiftory, in French.

#### SECTION 4.

After paffing the ftudies already fpecified, the Students in the Philofophical Clafs, before their admiffion to the firft degree in the arts, fhall be acquainted with Natural Philofophy and Aftronomy, Moral Philofophy, Conftitution of the United States of America, and of the different States, Logic, Metaphyfics, or the Philofophy of the Human Mind, Longinus and Horace's Art of Poetry; or, inftead of the Greek, with Buffon's Natural Hiftory, in French, or fome other approved French author.

# APPENDIX B

## UNION COLLEGE CURRICULUM of 1827

Jonathan Pearson's scrapbook contains the program required of all students one year before introduction of the Scientific Course.

### Course of Studies in Union College.

**FRESHMAN CLASS.**

1st Session.
- Cicero De Officiis.*
- Horace and Latin Prosody.
- Xenophon.*

2d Session.
- Cicero De Officiis, con. & Roman Antiquities, . . Adam's.
- Horace, continued, and a review of Virgil.
- Xenophon, continued.

3d Session.
- Roman Antiquities, continued.
- Xenophon, continued, and a review of the Greek Testament.
- Arithmetic, . . . . . , . . Pike's.

**SOPHOMORE CLASS.**

1st Session.
- Euclid's Elements of Geometry, . . . Simpson's.
- Geography.
- Cicero De Oratore.

2d Session.
- Euclid, continued.
- Collectanea Græca Minora.
- Cicero De Oratore, continued.

3d Session.
- Conic Sections, . . . . . Simpson's.
- Longinus.
- Logic, . . . . . . . Watts'.

**JUNIOR CLASS.**

1st Session.
- Algebra, } . . . . Day's.
- Fluxions, } . . . . Hutton's.
- Trigonometry and Spherics, . { Printed for the use of the Junior Students in U. College.
- Tacitus.

2d Session.
- Belles Lettres, . . . . Blair's.
- Mensuration, Guaging, } Flint's or Gibson's Surveying, Ewing's,
- Surveying and Navigation, } Moore's or Bowditch's Nav.
- Tacitus, continued.

3d Session.
- Natural Philosophy, . . . : . Enfield's.
- Homer.
- Moral Philosophy, . . . . . Paley's.

**SENIOR CLASS.**

1st Session.
- Chemistry, . . . . . . Henry's.
- Natural Philosophy, continued.
- Elements of Criticism, . . . . Paley's.

2d Session.
- Chemistry, continued, and Natural History, . Smellie's.
- Kaimes, continued, and Metaphysics, . . Stewart's.
- Astronomy, . . . . Enfield's or Furguson's.

3d Session.
- Butler's Analogy, Blackstone's Com. Greek & Heb. Languages.
- Lectures on Oratory, Chemistry and Natural Philosophy.

*The Grammars used in College are Moore's Greek Grammar, by Dr. Blatchford, and Rigg's Latin Grammar.*

## UNION COLLEGE CURRICULUM of 1828

The catalog announced Union's radical new program which allowed students to choose between the traditional Classical Course and the new Scientific Course which contained no classics after the common freshman year. Graduates of both courses received the B.A. degree.

# COURSE OF STUDIES,

## Preparatory for Admission into Union College.

---

Riggs' Latin Grammar, and Farrand's Latin Course.
Selectæ e Vet. Eutropius and Clark's Introduction.
Corn's Nepos, Cæsar's Commentaries.
Virgil, Cicero's Select Orations, Moore's Greek Grammar, Greek Testament, Greek Introduction and Græca Minora.
Arithmetic, English Grammar and Geography.
Less attention is paid to the particular books read, than to the amount of knowledge acquired.

---

# COURSE OF STUDIES,

## PURSUED AFTER ADMISSION INTO UNION COLLEGE.

# FRESHMAN CLASS.

### First Term.

Cicero de Officiis, de Amicit. &c.
Horace and Latin Prosody—*with Composition and Declamation.*
Herodotus and Thucydides, - - - - - - *Græca Maj.*

### Second Term.

Xenophon's Cyrop'a. . . . . . . *Græca Maj.*
Horace, Roman Antiquities.
Livy—*with Composition and Declamation.*

### Third Term.

Sallust.
Xenophon's Cyrop'a. and Anabasis.
Lysias, Isocrates and Demosthenes—*with Composition and Declamation.*

# CLASSICAL COURSE.*

## Sophomore Class.

### FIRST TERM.

Tacitus' History.
Xenophon's Mem. and Plato, - - - - - *Græca Maj.*
Logic.

### SECOND TERM.

Algebra—1st vol. Euler.
Aristotle, Dyonisius and Longinus, - - - - *Græca Maj.*
Tacitus, continued.

### THIRD TERM.

Jameison's Rhetoric.
Plane Geometry, - - - - - - - - *Legendre.*
Homer's Odyssey, - - - - - - - *Græca Maj.*

## Junior Class.

### FIRST TERM.

Blair's Lectures.
Plane and Spherical Trigonometry—with the Applications, - *Day.*
Hesiod and Sophocles, - - - - - - *Græca Maj.*

### SECOND TERM.

Cicero de Oratore.
Natural Philosophy, - - - - - - *Farrar's Mechanics.*
Conic Sections, - - - - - - - *West.*

### THIRD TERM.

Political Economy.
Medea, &c. - - - - - - - - *Græca Maj.*
Natural Philosophy, - - - - - - *Farrar's Mechanics.*

## Senior Class.

### FIRST TERM.

Intellectual Philosophy, - - - - - *Stewart or Brown.*
Natural Philosophy, continued—Optics, &c. - - - - *Biot.*
Elements of Criticism, - - - - - - *Kames.*

### SECOND TERM.

Astronomy, - - - - - - - - - *Biot.*
Moral Philosophy, - - - - - - - *Paley.*
Kames and Lectures on Chemistry.

### THIRD TERM.

Hebrew.
Hebrew.
Lectures on Elements of Criticism, Chemistry, Botany and Mineralogy.

---

# SCIENTIFIC COURSE.*
## Sophomore Class.

### FIRST TERM.

Tytler's History.
Arithmetic,    ·  -  ·  -  ·  -  ·  -  ·    *Hasler.*
Logic.

### SECOND TERM.

Algebra—1st vol. Euler.
Natural Theology,   -  ·  -  ·  -  ·  -  ·   *Paley.*
Natural History,   -  ·  -  ·  -  ·  -  ·   *Ware.*

### THIRD TERM.

Jameison's Rhetoric.
Plane Geometry,   -  ·  -  ·  -  ·  -   *Legendre.*
French or Spanish.

---

## Junior Class.

### FIRST TERM.

Blair's Lectures.
Solid Geometry,   -  ·  -  ·  -  ·  -   *Legendre.*
Algebra,   -  ·  -  ·  -  ·  -  ·   *Lacroix.*

### SECOND TERM.

Plane and Spherical Trigonometry, and the Applications,    *Hasler.*
Natural Philosophy,   -  ·  -  ·  -  *Farrar's Mechanics.*
Descriptive Geometry and the Applications,   -  ·  -  *Davies.*

### THIRD TERM.

Analytic Geometry,   -  ·  -  ·  -  ·  -   *Biot.*
French or Spanish.
Natural Philosophy,   -  ·  -  ·  -  *Farrar's Mechanics.*

---

## Senior Class.

### FIRST TERM.

Differential and Integral Calculus,   -  ·  -  ·  *Boucharlot.*
Natural Philosophy, continued, Optics, &c.   -  ·  -   *Biot.*
Elements of Criticism,   -  ·  -  ·  -  ·   *Kames.*

### SECOND TERM.

Astronomy,   -  ·  -  ·  ·  -  ·   *Biot.*
Moral Philosophy,   -  ·  -  ·  -  ·   *Paley.*
Kames and Lectures on Chemistry.

### THIRD TERM.

Anatomy or Blackstone.
Physiology or Kent.
Lectures on Elements of Criticism, Chemistry, Botany and Mineralogy.

# APPENDIX D

## UNION COLLEGE SCIENTIFIC COLLECTIONS IN 1876

In preparation for the nation-wide centennial celebrations of the United States, a proposal was made by Franklin Benjamin Hough, a prominent alumnus (1843), that a series of circulars be issued by the federal Bureau of Education giving historical and statistical data for American colleges and universities.

As a sample of the projected work, Hough prepared a booklet of 81 pages, giving a detailed "Historical Sketch" of Union College. His cover and four pages are reproduced directly from that publication.

HISTORICAL SKETCH

OF

# UNION COLLEGE

[ NOW A BRANCH OF UNION UNIVERSITY. ]

FOUNDED AT SCHENECTADY, N. Y., FEBRUARY 25, 1795.

———————

*Prepared in compliance with an invitation from the Commissioner of the Bureau of Education, representing the Department of the Interior in matters relating to the National Centennial of 1876.*

———————

WASHINGTON:
GOVERNMENT PRINTING OFFICE.
1876.

## COLLECTIONS IN NATURAL HISTORY.

From an early period the college has been a center of interest for students of natural history, and collections were added from time to time, especially in 1841, when a considerable number of minerals and fossils were received from the State cabinet. In 1860 the "Wheatley collection" of shells and minerals, valued at the time as worth $20,000, and now still more, was presented by Mr. E. C. Delavan.

The dredgings upon our coast in recent years have enriched the cabinet with many forms of marine life, and within the last three years an extensive collection of specimens was added by Prof. H. E. Webster as the result of his labors in dredging at Eastport, Me., on the coast of Massachusetts and Virginia, and the west coast of Florida.

A valuable herbarium has been given by Dr. George T. Stevens, of Albany.

## PHILOSOPHICAL DEPARTMENT.

In this department the collections, under the care of Prof. John Foster, have grown to be among the finest in the country. The donations of friends have added largely to their value,† but the principal part has been purchased by the friends of the college or by special funds raised for this pur-

---

\* In 1873 Mr. James Brown, of New York, gave the sum of $10,000, under the name of the "Coe memorial fund." The income is applied to increasing the college library, which it does at the rate of about two hundred volumes a year.

† The donors to this department are William H. H. Moore, Hon. A. H. Rice, Henry C. Potter, M.D., Henry R. Pierson, Howard Potter, William A. Whitbeck, C. N. Potter, Lemon Thomson, and A. Q. Stevens.

pose. The professor has very recently, while in Europe, procured many articles of especial interest as illustrating the more advanced discoveries of the day. The more important instruments owed by the college are as follows:

IN ELECTRICITY: Thompson's divided ring electrometer and reflecting galvanometer; Wheatstone's bridge; British Association standard unit of resistance; positive and negative electrophorus; Holtz machine, by Ruhmkorff; Grove's galvanic battery of 40 elements; small induction coil, giving spark of 2 inches, by Ruhmkorff; large coil with interrupter giving spark of 17 inches; battery of 4 jars, *en cascade*, for the large coil; Chester battery of 8 large elements; Bunsen galvanic battery of 60 elements; Foucault's electric lamp; collection of Geissler tubes; magneto-electric machine; Morse register and relay magnet; Gaugain's tangent compass; Lamont's electrometer for atmospheric electricity; pile of Zamboni; large thermo-electric pile of 36 elements according to Marcus's method of construction; jar with movable coatings; apparatus for piercing glass with electricity.

IN MAGNETISM: Lamont's magnetic theodolite for determining the absolute intensity; additions to the theodolite for finding the absolute declination; dipping needle for observations; magnetic engines.

IN LIGHT: Porte lumière; Duboscq's magic-lantern, adapted to the use of either the electric or lime light; Marcey's sciopticon; complete photographic apparatus; circle for demonstrating the laws of reflection, refraction, polarization, &c.; Duboscq's apparatus for projecting upon a screen all the phenomena of double refraction and polarization; solar microscope with collection of objects; prism for the limiting angle; equilateral flint-glass prism; hollow prism with compartments for different liquids; polyprism; mounted achromatic lens; 3 bisulphide of carbon lenses; total reflection fountain; spectroscopes.

IN HEAT: Ruhmkorft's thermo-electric multiplier and pile; line pile for showing calorific spectrum; collection of plates for diathermancy; apparatus of Despretz for conduction; apparatus of Gay-Lussac for tension of vapors; apparatus of Senarmont for the conduction of heat in crystals; thermometer with reservoir; weight thermometer; wet bulb hygrometer; Breguet's metallic thermometer; differential thermometer; apparatus of Tralles for maximum density of water; set of balls of different metals for specific heat; fire syringes of brass and of glass; Regnault's hypsometer and hygrometer; Wollaston's eryophorus.

IN ACOUSTICS: From König of Paris: mouth-pieces of several instruments; model of locomotive whistle; set of 10 diapasons with resonant cases; set of 19 Helmholtz resonators; double sirene of Helmholtz; 5 diapasons with resonators for the vowel sounds; large soufflerie for organ-pipes and sirene; 64 organ-pipes for demonstrating theory of vibrating air columns; sonometer with 8 cords; apparatus of Melde for vibrating cords; König's new apparatus for interference, shown by manometric flames; sets of plates for acoustic figures; ear and speaking trumpets; Dr. Auzoux's models of the ear and the larynx; wire-coil for showing the mode in which both light and sound waves are propagated, presented by Blake Brothers, of New Haven, Conn.; apparatus of Lissajou for showing vibrations by both the optical and graphical methods; Wheatstone's kaleidophone and wave apparatus; Schaffgotsch's apparatus; Quincke's apparatus for measuring wave length.

IN PNEUMATICS: Air-pumps with their apparatus; Magdeburg hemispheres and planes; apparatus for compressing air; apparatus for proving Mariotte's law.

IN STATICS AND DYNAMICS: Mechanical powers; Atwood's machine; whirling-table; pendulum, &c.

IN HYDROSTATICS AND HYDRAULICS: Hydrostatic bellows and press; hydrometers; Pascal's vases; Mariotte's flask; Prony's floater; apparatus for demonstrating the laws of spouting fluids; models of different forms of fountains; hydraulic ram; models of various forms of pumps; models of water-wheels.

FOR PRECISE MEASUREMENTS: Steel scales of English and French measures; graduated vessels of various measures and volumes; balances by Becker and other makers; spherometer by Buff and Berger; Wollaston's goniometer; theodolites, &c.

The following are some of the principal pieces recently received from London and Paris:

*Mechanics.*—Inclined plane of Galileo, Atwood's machine; apparatus of Bourdon Kater's pendulum; manometer of Bourdon; hydrostatic balance; gyroscope of Hardy and of Fessel; models of screws, of pendulums, and of escapements; large apparatus showing the principal transformations of movement-dividing machine.

*Heat.*—Volumometer of Say and of Regnault; pyrometer with dial; apparatus for the absolute dilatation of liquids; of Regnault, for dilatation of gases, both under constant pressure and under constant volume; of Dalton for tension of vapors; of Regnault for same; of Gay-Lussac

4 U

APPENDIX D (continued)

for tension of vapors below freezing-point ; of Dalton for same in a vacuum ; of Gay-Lussac for tension of mixed vapors and gases ; of Dumas for density of vapors ; of Gay-Lussac for same ; of Regnault, with manometer, for density of gas ; of Ritchie for emission and absorption of heat ; of Jamin for the conduction of heat ; calorimeter of Lavoisier and Laplace ; of Regnault for specific heat by cooling ; of same, large size ; for specific heat by mixtures ; of Favre and Silbermann for measuring heat of combustion ; of Dupretz for measuring latent heat of vaporization ; of Regnault, for measuring the elastic force of compressed air, and also of the tension of vapors both above and below 100 degrees C., complete ; pyrheliometer of Pouillet ; cathetometer, one meter in length, graduated ; apparatus of Pouillet for measuring the compressibility of gases ; of Simon for capillarity ; of Bouligny, for the spheroidal state, complete ; set of lenses, prisms, plates, &c., for the Melloni apparatus ; large machine of Natterer for liquefaction of azote and of carbonic acid ; Carré's air-pump, exhausting and condensing with sulphuric acid ; reservoir.

*Electricity and magnetism.*—Large reflecting galvanometer of Weber with telescope ; vertical differential galvanometer ; apparatus of Ampère ; electric planisphere ; gas pile of Grove ; secondary pile of Planté ; large magnet of Jamin ; Alliance magnets ; electric machine ; portable Grove's battery of fifty elements for the electric light ; large electro-magnet for diamagnetism, rotation of polarized ray, &c. ; Delezenne's circle ; thermo-electric pile ; set of resistance coils with bridge.

*Acoustics.*—Regnault's chronograph with clock ; apparatus of Crovee for projection of wave motions.

*Light.*—Large heliostat of Silbermann ; several forms of apparatus for projecting colored rings of thin plates ; interference refractor of Jamin ; circle, complete, of Jamin and Senarmont ; combined polariscope and polarizing microscope ; Biot's apparatus, complete, for rotary polarization, including Soleil's saccharometer ; vertical lantern ; prism of Dessains ; large prism of Foucault ; polariscope of Arago ; photometer of Foucault ; Becquerel's phosphoroscope ; prisms of Senarmont, of Hartnack, of Jansen, of Rochon, and of Silbermann ; apparatus of Delezenne ; of Stokes ; of conical prisms for caustics by reflection ; of parabolic mirrors ; of seven mirrors for recomposition of light ; of two large piles of glass for polarization ; a large Steinheil spectroscope with four prisms ; large telemeter of Gautier.

*Measurements.*—Sets of French weights and measures of length and of capacity, dry and fluid.

# APPENDIX E

## NATURAL PHILOSOPHY & PHYSICS FACULTY AT UNION COLLEGE, 1795–1895

Colonel John Taylor: 1795-1801

Cornelius Van den Heuvell: 1798-99

Benjamin Allen: 1800-1809

Ferdinand Rudolf Hassler: 1810-1811

Thomas Macauley: 1811-1822

Alonzo Potter: 1822-1826

Francis Wayland: 1826-1827

Benjamin Joslin: 1827-1837

Isaac Jackson: 1826-1877

John Foster: 1836-1885

Isaiah Price: 1878-1884

Thomas W. Wright: 1885-1905

Howard Opdyke: 1894-1928

# APPENDIX F

## PARTIAL LIST OF 19TH CENTURY UNION GRADUATES
## WITH NOTABLE CAREERS IN SCIENCE OR SCIENCE EDUCATION

Information compiled from Raymond's *History of Union College, Vol. III; Dictionary of American Biography; Dictionary of Scientific Biography* and College Archives.

**Class:**

1803    **Bowen, William C.** (M.D. Edinburgh, 1807) Physician & Chemist.
First Professor of Chemistry, Brown Univ. Medical School, 1811-13.

1804    **Brownell, Thomas.** Classicist, Chemist & Clergyman.
Tutor, Union College, 1805-06.
Professor of Logic & Belles-Lettres, Union, 1806-11.
Lecturer on Chemistry, Union, 1811-14.
Professor of Rhetoric & Chemistry, Union, 1814-19.
A founder and first President of Trinity College.

1804    **Macauley, Thomas.** Scientist & Clergyman.
Tutor, Union College, 1805-11.
Lecturer on Mathematics & Natural Philosophy, Union, 1811-14.
Professor of Mathematics & Natural Philosophy, Union, 1814-22.
First President of Union Theological Seminary.

1807    **Beck, T. Romeyn.** (M.D., 1811.) Physician, Scientist, Educator, Author.
Private medical practice, Albany, 1811-17.
Principal of Albany Academy, 1817-53, & mentor to Joseph Henry.
First Vice President of R.P.I.
Lecturer on Medical Jurisprudence & Professor of the Institutes of Medicine, Western College of Physicians & Surgeons, Fairfield, N.Y., 1815-40.
Professor of Materia Medica, Albany Medical College, 1840-53.
Author of *Elements of Medical Jurisprudence, 1823.*

1813    **Wayland, Francis.** Educator & Author.
Tutor, Union College, 1817-21.
Professor of Natural Philosophy, Union, 1826-27.
President of Brown University, 1827-55, where he introduced many of Nott's curricular reforms.
Author of *The System of Collegiate Education in Our Age and Country.*

1817    **Beck, Lewis C.** (M.D.) Physician, Chemist, Botanist & Geologist.
Brother of T. Romeyn Beck
Teacher of Botany, Berkshire Medical Inst., 1824.
Professor of Botany, Zoology & Mineralogy at R.P.I., 1824-25.
Professor of Botany & Chemistry, Vermont Academy of Medicine, 1826.
Professor of Chemistry & Natural History, Rutgers, 1830-53.
Professor, Albany Medical College, 1840-? (simultaneous with Rutgers)
Author of many publications; in 1827 he collaborated with Joseph Henry in producing a Scale of Chemical Equivalents.
Author of *A Manual of Chemistry, 1831.*

1817    **Dexter, George.** Albany Druggist. Manufacturer & Importer of anatomical models and philosophical apparatus.

1817    **Nott, Joel.** Chemist.
Tutor in Chemistry, Union College, 1820-22.
Professor of Chemistry, Union, 1822-31.
Author of *A Syllabus of Lectures on Chemistry.*

1818     **Potter, Alonzo.**   Clergyman & Educator.
Tutor & Professor of Natural Philosophy at Union College, 1819-26.
Vice President, Union, 1831-45.
Protestant Episcopal Bishop of Pennsylvania, 1845-65.

1821     **Joslin, Benjamin F**. (M.D.) Physician, Scientist & Educator.
Principal of Schenectady Academy, 1821-22.
Tutor at Union College, 1822-24.
Professor of Chemistry & Natural Sciences at Polytechnic, Chittenango, 1826.
Professor of Mathematics & Natural Philosophy at Union, 1827-37.
Professor of Mathematics & Natural Philosophy, C.C.N.Y., 1838-44.
Author of two widely-used books on homeopathy and many journal articles
    on original observations in physics, meteorology, mechanics, homeopathy
    and medicine.
Assisted Stephen Alexander in observation of artificial star occultations.
Author of *Meteorological Observations & Essays.*
Editor of Lardner's *A Treatise on Hydrostatics and Pneumatics.*

1824     **Alexander, Stephen.**   Astronomer and Mathematician.
Teacher, Chittenango Academy, 1825-30.
Teacher, Albany Academy, 1830-33.
Tutor at Princeton, 1833.
Adjunct Professor of Mathematics, Princeton, 1834-40.
Professor of Astronomy, Princeton, 1840-78 & Professor of
    Mathematics, 1845-54.
Chief of Expedition to Labrador to observe solar eclipse, Aug. 1869.
Worked on many projects with Joseph Henry at Albany and Princeton.
Author of a text on mathematics and many scientific papers on mathematics
    and astronomy.

1825     **Gale, Leonard D.**   (M.D.) Physician & Scientist.
Assistant Professor, then Professor of Chemistry, New York College of
    Pharmacy, 1830-34.
Professor of Chemistry, Geology and Mineralogy, C.C.N.Y., 1834-40.
Conducted Geological Survey of Manhattan, 1833.
Perfected electromagnetic apparatus for Samuel F.B. Morse's telegraph, 1837,
    utilizing recent discoveries on magnetic induction by Joseph Henry.
Assistant Examiner, Patent Office, Washington, D.C., 1846-48.
Chief Examiner in Chemistry, Patent Office, 1848-57.
Author of *Elements of Chemistry* (1835) & *Elements of Natural
    Philosophy* (1837).

1825     **Tappan, Henry.**   Clergyman & Educator.
First Professor appointed by C.C.N.Y., 1832.
President of the Univ. of Michigan, 1852-63, where he pioneered in the
    introduction of graduate studies in science.

1826     **Jackson, Isaac W**. Scientist & Educator.
Spent 50 years ( his entire professional life) at Union College teaching
    natural philosophy, physics and mathematics.
Author of texts in Mechanics, Optics, Conic Sections & Trigonometry.

1826     **Potter, Horatio.**   (brother of Alonzo). Clergyman & Scientist.
Professor of Natural Philosophy at Trinity College, 1828-33.
Episcopal Bishop of New York, 1861-87.

1828    **Averill, Chester**. Chemist.
Professor of Botany & Chemistry, Union College, 1831-36.

1830    **Stewart, Duncan L.** Educator.
Professor of Mathematics & Natural Philosophy, Trinity College, 1837-41.

1830    **Totten, Silas S.** Scientist & Educator.
Professor of Mathematics & Natural Philosophy, Trinity College, 1833-37.
President of Trinity, 1837-48.
Chancellor of University of Iowa, 1860-62.

1830    **Whipple, Squire**. Inventor & Civil Engineer.
Surveyor for Baltimore and Ohio Railroad & Erie Canal & New York and
  Erie Railroad.
Inventor of a Protracting Trigonometer, 1833.
Constructed some 25 leveling instruments, transits & theodolites.
Patented the "Whipple Truss Bridge Support," 1841, a bowstring truss, made
  of cast and wrought iron. This design was adopted for all Erie Canal
  bridges, and Whipple became known as "Father of Iron Bridges."
Builder of many other road & rail bridges in New York State & the
  Midwest.
Author of *A Work Upon Bridge Building* (1847), that described, for the first
  time, the mathematical design of trusses.

1831    **Park, Roswell.** Soldier, Engineer & Educator.
Graduated first in class of USMA at West Point, same year as Union
  graduation.
Military engineer, 1831-36, constructing forts and dams.
Professor of Chemistry & Natural Philosophy, Univ. of Pennsylvania,
  1836-42.
Educator in several Midwestern prep schools and colleges, teaching scientific
  studies and special scientific programs.
Author of books and papers on poetry, history of the USMA, magnetism, and
  a survey of knowledge.

1832    **West, Charles E.** Pioneer in female education.
Principal of Rutgers Female Institute, 1839-51, where he introduced the first
  college studies for women in Advanced Algebra, Analytic Geometry &
  Calculus, as well as Chemistry and Natural History, with laboratories.
Principal of Buffalo Female Seminary, 1851-60.
Principal of Brooklyn Heights Seminary, 1860-89.
Founding member of American Assn. for the Advancement of Science.
Correspondent of Joseph Henry, to whom he provided laboratory facilities for
  experiments on atmospheric electricity in 1842.

1835    **Pearson, Jonathan.** Historian, Educator, Diarist, Author.
Tutor, Union College, 1836-39.
Adjunct Professor of Chemistry & Natural History, 1839-50.
Professor of Chemistry & Natural History, 1850-57.
Professor of Natural History, 1857-73.
Professor of Agriculture & Botany, 1873-87.
Union College Librarian & Treasurer.
Author of five books on Albany, Schenectady, Genealogy & the Dutch
  Reformed Church.

1835     **Foster, John**. Scientist, Educator & Author.
Tutor, 1836-39 at Union, where he spent his entire professional life.
Assistant Professor of Mathematics & Natural Philosophy, 1839-49.
Professor of Natural Philosophy, 1849-85.
Primarily responsible for Union's fine collection of 19th century Physics apparatus.
Author of a text on Electricity, Magnetism & Acoustics.

1836     **Stoddard, Orange N.** Scientist & Educator.
Professor of Natural Science, Hanover College, Indiana.
Vice President and Professor of Natural Science, Miami University, 1845-70.
Professor of Natural Science, Wooster University, Ohio, 1870-89.

1837     **Halleck, H. Wager.** Military Engineer & Commander, Author.
Acting Professor of Engineering, USMA at West Point, 1839-40.
Commander-in-chief, U.S. Army, 1862-64.

1837     **Perry, Stewart.** Businessman & Inventor.
Patented early models of a gas engine in 1844 and 1846.
Inventor of agricultural implements, locks, stereopticon, velocipede, etc.

1837     **Tuckerman, Edward Jr.** Attorney, Classicist & Botanist.
Professor of History & Botany, Amherst College, 1858-86.
Author of many scientific publications.

1840     **Morgan, Lewis A.** Ethnologist, Lawyer, Author.
"Father of American Ethnology."
President of American Academy for the Advancement of Science, 1880.

1843     **Hough, Franklin Benjamin.** (M.D. & Ph.D.) Surgeon, Botanist, Geologist, Meteorologist, Historian, Statistician, Civil Servant, Editor.
Noted for collections of rocks, birds and fossils, some of which were acquired by the Smithsonian Institution.
Author of four New York county histories (Lewis, Jefferson, Franklin and St. Lawrence), and thousands of reports on historical studies, indian treaties, legislation, archeology, meteorology, astronomy, forest fires, and other natural phenomena, & educational affairs for the Board of Regents.
Assistant Secretary of State to New York State, 1852–55; major organizer of and contributor to State Gazette in 1860 and 1873.
Superintendent of NYS Census in 1855 and 1865 and preparations for the federal census of 1870.
"Father of American Forestry," author of many articles and textbooks. First U.S. Forestry Agent, 1877-83.
Editor of *American Journal of Forestry*.
Helped establish the N.Y.S. Forest Preserve in 1885.

1844     **Joy, Charles A.** (Ph.D., Göttingen, 1853) Scientist & Educator.
Professor of Chemistry at Union College, 1855-57.
Professor of Chemistry, Columbia University, 1857-77.
Editor of *Scientific American and Journal of Applied Chemistry*.
President of Lyceum of Natural History (1886) & American Photographers' Society.

1846     **Veeder, Peter V.** Clergyman, Scientist & Educator.
Principal of City College, San Francisco, 1865-71.
Dean of the Faculty & Professor of Physics, Imperial University of Tokyo, 1871-72.
Professor of Mathematics, Western University (Pittsburgh) 1880-82.
Professor of Mathematics, Wake Forest University (N.C.), 1882-86.

1851 **Gurley, Lewis E.** Engineer, Manufacturer & Editor.
One of largest U.S. manufacturers of high-quality precision instruments for surveying and hydrology in partnership with brother (William) in Troy, N.Y.
Edited first & subsequent editions of "Manual of Instruments for Engineers and Surveyors," 1855-84.

1852 **Murray, David.** (Ph. D.) Mathematician, Scientist & Educator.
Principal, Albany Academy, 1857-63.
Professor of Mathematics & Astronomy, Rutgers University 1863-73.
Adviser to Japanese Imperial Minister of Education, 1873-79.
    Established a universal education system in Japan along American lines.

1853 **Potter, Edward Tuckerman.** Architect, Engineer & Composer.
Innovative architect & developer of new structural technology.
Designer of the Nott Memorial at Union College, with many unique features, as well as many churches and other edifices.

1854 **Miller, Prosper M.** Scientist & Educator.
Prof. of Natural Science, Alfred University, 1868-72.

1856 **Hough, George Washington.** Astronomer.
Astronomer, Dudley Observatory, 1860-62.
Director of Dudley Observatory, 1863-74.
Director of Dearborn Observatory & Professor of Astronomy, Univ. of Chicago, 1879-87.
Professor of Astronomy, Northwestern University, 1887-1909 (?)
Particularly noted for studies of Jupiter, discovery of some 650 new double stars & invention of astronomical & meteorological instruments.

1856 **Norton, Sydney A.** (M.D. & Ph.D.) Physician, Chemist, Teacher & Author.
Tutor in Mathematics & Assistant in Chemistry, Union College, 1857.
Teacher of Natural Science at Cleveland High School & Mt. Auburn Seminary.
Professor of Chemistry, Miami Medical College, Ohio, 1869-73.
Acting Professor of Physics, Union College, 1873.
Professor of Chemistry, Ohio State Univ., 1873-1900 (?)
Author of two texts in physics, and two in chemistry.

1858 **Cooley, Leroy C.** (Ph.D.) Author and Educator.
Professor of Natural Science, NYS Normal College, 1861-74.
Professor of Physics and Chemistry at Vassar, 1874-93.
Author of four physics textbooks and three in chemistry.

1858 **Strong, Edwin A.** Scientist & Educator.
Superintendent of Public Schools in Grand Rapids, Mich., 1862-71.
Professor of Physics at Michigan State Normal School, Ypsilanti, 1885-1916.
President of Michigan State Teachers Assn. and Kent Scientific Institution.

1862 **Chandler, William H.** (Ph.D.) Scientist & Educator.
Instructor in Chemistry, Columbia School of Mines, 1868-71.
Professor of Chemistry, Lehigh College, 1871-1906.
Co-editor, *American Chemist*, 1871-77.

1863 **Fearey, Thomas H.** Signal Corps officer.
Professor of Applied Physics, Vanderbilt University, 1887.

1865     **Staley, Cady**. Engineer, Author.
Professor of Civil Engineering, Union College, 1868-86.
Dean, Union College, 1876-86.
President of Case School of Applied Science, 1886-1902.
Author of three texts on structures.

1865     **Keep, William John.** Metallurgist.
Worked for Michigan Stove Company for many years.
Author of *History of Heating Apparatus.*
Consulted with Nott on latter's experimental model of Saracen Stove in 1865.

1872     **Price, Isaiah**. Civil Engineer, Mathematician, Physicist.
Tutor at Union College in Math and History, 1872-75.
Adjunct Professor of Physics, Union, 1876-78.
Professor of Mathematics and Assistant Professor of Physics, Union, 1878-84.

1874     **Hoadley, George A**. (A.M.) Civil Engineer & Author.
Professor of Physics, Swarthmore, 1883-94.
Vice President of Swarthmore, 1894-1914.
Author of four physics textbooks.

1890     **Lockner, Sydney J.** Astronomer & Mathematician.
Astronomer at Dudley Observatory, 1890-93.
Professor of Mathematics & Chemistry, University of Akron, 1912-20.
Professor of Mathematics, University of Pittsburgh, 1920-25.

1891     **Lay, William O.** Astronomer & Bridge Builder.
At Dudley Observatory, 1891-93.

1895     **Wright, William Howard**. Chemist.
Founder of Schenectady Varnish Company, later Schenectady Chemicals.

1899     **Wright, Frank Thomas**. Chemist.
Worked for Westinghouse and for U.S. Cast Iron Pipe Co.

# APPENDIX G

## PHYSICS AT UNION COLLEGE FROM 1895 TO 1995

By David Peak
Frank and Marie Louise Bailey Professor of Physics

July 17, 1994

In the earliest years of the 20th century, the department was referred to as "Physics and Mechanics." Indeed, its curriculum during this period appears to have been directed to providing aspiring engineers with a firm understanding of the mechanical properties of matter. A perusal of students' class notes from these years reveals an introductory physics course that concentrated on the mechanics of fluids, thermodynamics, and geometrical optics. A smattering of calculus permeated the presentation. Though features of the magnetic field were discussed, the electric field was essentially absent, and so, obviously, was any reference to Maxwell's equations. There was no mention of the atomic structure of matter nor of relativity—those notions having then only recently become widely accepted. Nonetheless, introductory physics at Union in the early 1900's still had the feel of modernity: for example, one finds no reference to such outmoded topics as the mechanical advantage of levers and pulleys.

The primary physics instructor through the first two decades of the new century was Howard Opdyke, who taught at the College from 1894 to 1928. Opdyke was a careful and stylish dresser and an avid follower of Union athletics. By the first decade of the new century, Opdyke had attained considerable stature within the faculty, and a reputation for eccentricity among the students. Ralph Bennett ('21) recalls: "Oppie's lectures were very elegant performances. In the one we all remember he was faced with the problem of 'picking a point at random' on the blackboard. To make sure his selection was indeed random he pulled his elegant silk handkerchief from his breast pocket, and wrapped it around his head to blindfold himself. Then, chalk in hand, he turned around three times, touched the chalk to what he thought was the blackboard, but walked through the open door into the laboratory, first to the concern and then to the amusement of us all."

Joining Opdyke in 1915 was Richard Kleeman, a figure of some note. The May 1915 issue of the *Union Alumni Monthly* cites a visit by Kleeman to Schenectady as "a great compliment," claiming him to be "one of a half dozen of the world's greatest physicists." Kleeman had done work with Nobel laureates W. H. Bragg and J. J. Thompson and had written numerous papers and several books, most of which were devoted to what now would be called chemical physics. Though today we see the *Monthly's* accolade as hyperbole, Kleeman did produce a substantial body of research over the years, some even with Union undergraduates.

Kleeman's other-worldliness left him vulnerable to sophomoric pranks in the classroom. Though Kleeman's teaching assignment was soon restricted to laboratory instruction solely, that wasn't totally effective in protecting him. Bennett remembers, for example, how students in the lab learned to blow air back into the gas line, sometimes extinguishing a bank of four or five Bunsen burners simultaneously and leaving Kleeman hopelessly befuddled. (Fortunately, Kleeman's antagonists never achieved an explosive gas-air mixture.)

In 1919, the well known physicist Floyd Richtmyer came to Schenectady for two years, on leave from his permanent position at Cornell, to teach at Union and to do research at the General Electric research laboratory. Richtmyer is well-known among older readers for his watershed text on "modern physics" ("Richtmyer and Kennard")—dealing with relativity, early quantum theory, and the nature of the atom—and his appearance at Union marked a symbolic transition from the macroscopic physics of the pre-twentieth century classical world to the microscopic picture upon which the contemporary view of physical reality is based.

This transition was solidified by the hiring of Peter Wold as professor and chair of physics in 1920. Wold had been a research physicist at the Western Electric Company, studying the newly emerging technology of vacuum tubes. His interests meshed well with work going on at General Electric and for many years Wold nurtured a close relationship between GE and Union. One tangible aspect of this relationship was that, from 1928 until the onset of World War II, the heralded GE scientists Irving Langmuir and Albert Hull held joint appointments in physics, the former lecturing on vacuum tube phenomena, the latter on the production of x-rays and x-ray crystallography. Prior to coming to Union, Wold had already established an international reputation in education. In the years preceding World War I he was instrumental in helping the Chinese government transform its ancient educational system to one more closely mirroring that of the United States. At Union, Wold rapidly gained respect as a charismatic and engaging lecturer and an educational visionary. Many of the students of his department cite his compelling lectures in introductory physics as convincing them to change their intended majors to physics. He introduced the B.S. in Physics degree in 1922, guided by his perception that a modern technological society would have great need for generalists educated in the fundamentals of science and mathematics. This degree typically required 18 to 20 credit hours per term (when 15 was the requirement for other degrees), including four full years of chemistry and mathematics along with an even heavier dose of physics. Two terms of research were also required. Despite its extraordinary demands, the B.S. in Physics was attained by 3 to 5 majors per year for many years. During the early 1920's the groundwork was also laid for a small master's degree program, and graduate courses in mathematical physics, electromagnetism, and quantum mechanics began to be offered on a regular basis.

In 1930 Wold made two very important hires: Vladimir Rojansky (a theorist) and Frank Studer (an experimentalist). For the next 15 years, Wold, Rojansky, and Studer formed the nucleus of one of the strongest small physics departments in the country. Among the students they produced were a president of GTE and long-time Union College trustee (Lee Davenport, '37), a recognized co-inventor of the laser (R. Gordon Gould, '41), a Nobel prize winner in medicine (Baruch Blumberg, '45), and many others who achieved high levels of distinction in academia and industry.

Rojansky is a legendary character. As a young man in the Russian White Army, he escaped capture by the Red Army by fleeing across Siberia on horseback. After immigrating to America he became the Ph.D. student of John Van Vleck. A master of the new quantum theory, Rojansky wrote an expository text on the subject that eventually became a standard around the world and brought international recognition to Union for decades. Rojansky was a superb lucid lecturer and a fabled storyteller, and he wielded a powerful intellectual influence over all students with whom he came in contact.

By his own self-effacing recollection, Studer was, in comparison "just conventional." Studer's reminiscences of Union College in the 1930's and early 1940's have a nostalgic charm. On gentility lost: Shortly after he and Rojansky arrived on campus, they were invited to the Wold's for "dinner at 7." As the day and time approached they (fortunately) learned that "dinner at 7" at the College meant formal attire, a revelation that sent them scurrying just in the nick of time to Richmans to buy tuxedos and black ties. On dollars and the deity: Wold felt that Philosophical Hall was too small and antiquated for a modern physics program and spent much effort raising funds for renovation. (Indeed, he succeeded in obtaining very large gifts from GE and the American Locomotive Company for this and associated purposes.) One contact was a wealthy widow with whom he had had many lengthy discussions. Finally, one day the widow came to Wold's office to conclude a deal. Before doing so, however, she asked if they could kneel together to ask for divine guidance. Later, when Wold was asked what he was praying for at that moment he replied, "that nobody would come into the office!" On the dangers of avuncular advice: As World War II was concluding, Studer was the only member of the old department teaching on campus. The class of 1945 included many very talented young men. One by one, they came to Studer to ask, "What should I do with the rest of my life?" To each, Studer snapped, "Go to graduate school," then patiently explained what that was and how they could finance it. Finally, Baruch Blumberg, later to receive advanced degrees in physics, mathematics, medicine, and, a vast collection of awards and honors in addition to his Nobel prize, came to seek Studer's counsel. Without hesitation, Studer told him, "Get a job."

Despite heavy teaching loads, Wold, Rojansky, and Studer managed to keep active as professional physicists. The three, for example, were central in establishing the New York Section of the American Physical Society. The inaugural meeting of the State Section occurred at Union on April 2, 1938 and Wold was the Section's first chairman.

The War virtually decimated the department. Wold died in 1945 and Studer left the College to do research at GE. The responsibility for rebuilding the program and guiding the department fell to Rojansky. He managed to bring to the College two professors—Ted Goble and Win Schwarz—who would be integral parts of the new department for many years to follow. But, administration was neither Rojansky's interest nor his strength. In 1948, Harold Way was brought in as the new chairman.

Way had considerable prior experience as a chair and as acting president at Knox College, and, with the backing of President Carter Davidson, his longtime friend from Knox, he concluded the reconstruction Rojansky had started. Among Way's hires who eventually became tenured at Union were Charles Swartz, Curt Hemenway, Robert Vought, Kenneth Baker (later, the president of Harvey Mudd College), Ennis Pilcher, Richard Henry ('54), Kenneth Schick, and Frank Titus.

The Frank and Marie Louise Bailey Professorship in Physics was established in 1949 to honor Rojansky—and to help him pay for the rapidly accelerating medical expenses of his wife. The latter situation continued to worsen and eventually, in 1955, it forced Rojansky to leave the College for a much higher paying position with the Ramo-

Woolridge Company (now TRW) in California. (Subsequently, after his retirement from industry, Rojansky became professor of physics at Harvey Mudd, uniting there once again with Baker.) In 1959, Way was installed as the second Bailey Professor.

Starting in the early 1950's, Rojansky and Way cooperated with Saul Dushman of the General Electric Company to institute a program of summer study for high school teachers that became the prototype for the very successful Summer Science Fellowship of the National Science Foundation.

Bolstered by hordes of returning veterans, a booming technological postwar economy, and elevated levels of federal support for research, the number of graduates in physics at Union grew rapidly in the 1950's and 1960's, reaching a high of 33. Despite the College's elimination of the demanding B.S. in Physics degree in 1951, students of Way's department remained highly attractive to the nation's graduate programs. These students obtained Ph.D.'s at a high rate (in 1960-61, for example, only nine other colleges or universities in America had more of its bachelor degree holders receive Ph.D.'s in physics) and many subsequently carved out meritorious careers. Their professional recognitions include the Wetherill Medal of the Franklin Institute (Frank Stern, '49), the R. W. Wood Prize of the American Optical Society (Sven Hartmann, '54), the E.O. Lawrence Award of the U.S. Atomic Energy Commission (Robert Walker, '50), and the Wolf Prize conferred by the Israeli government (Martin Perl, no degree). (Perl was a chemical engineer working at GE in 1950 when he began taking courses in math and physics at Union. According to Perl, Rojansky's lectures were so compelling he was left with no choice but to resign from GE and pursue an advanced degree in physics.)

The Union physics faculty produced a number of texts in the 1960's—Schwarz on electromagnetism, Goble and Baker on modern physics, and Henry on electronics—all of which enjoyed significant popularity for extended periods.

Way's retirement in 1966 marked the end of the long tradition of a monolithic chairmanship. Thereafter, administration of the department cycled internally among faculty already on staff: from Goble (1966), to Schick (1970), to Pilcher (1978), to Titus (1980), to David Peak (1984), to Gary Reich (1990), to Jay Newman (1992) (the latter three, along with Chris Jones, and most recently, Seyfi Maleki, becoming tenured members of the faculty in the post-Way years). The Bailey chair, vacant after Way's retirement, was filled from 1976-79 jointly by Schwarz and Swartz, and then again by Peak and Schick starting in 1987.

In the 1970's and 1980's, external events strongly influenced the direction of the department: federal funding for science fell sharply from the days just after the launch of Sputnik and the job market for physicists nose-dived as well. Undergraduate enrollments in physics at Union in the latter decades dropped back to pre-World War II levels and the never large master's program was discontinued. Despite their reduced numbers, students remained generally interested in pursuing advanced study. A much keener emphasis on scholarly work began to pervade the College and the faculty aggressively built up a research capability rarely encountered in physics at a small institution—with laboratories in light scattering, spectroscopy, particle detection, solid state, nonlinear dynamics, and accelerator physics. In 1973, Herb Strong, the inventor of the artificial diamond, joined the faculty as a research associate, and in 1986, Ralph Alpher internationally renowned for his seminal work on the Big Bang model of the universe, was named Distinguished Research Professor.

The tradition of textbook writing continued through the 1980's and 1990's. A widely adopted introductory textbook by Hans Ohanian, a faculty member for six years in the 1970's and 1980's, appeared in 1985 and a text for the general reader on the principles and applications of nonlinear dynamics by Peak was published in 1994.

By the College's Bicentennial, the Physics Department could reflect on a remarkable legacy: in addition to their many awards, its students had become directors of several academic and industrial laboratories, had become faculty members at over 50 colleges and universities, and had attained the honor of being named Fellow of the American Physical Society at a rate twice that of the Society's general membership. Over the span from 1920 to 1990, among the roughly 900 small liberal arts colleges in America, Union ranked 6th in the number of its graduates who received a Ph.D. in physics.

Sources: 1. Private communications from and personal interviews with Frank Studer and Harold Way. 2. Private communications from Ralph D. Bennett, '21, George A. Campbell, '32, Thomas J.Dietz, '32, Robert J. Doig, '38, Everett M. Haffner, '40, Harry H. Hall, '26, Clement L. Henshaw, '28, Louis C. Maples, '38, Harold L. Saxton, '24, J. Carl Seddon, '33, and Alford E. Stafford, '29. 3. *Union Alumni Monthly,* May, 1915. 4. *Concordiensis,* Vol. 38, No. 24, May 6, 1915. 5. "Three More Early Graduates of Distinction," V.A. Edgeloe, *Journal of the Historical Society of South Australia,* Number 13, pages 100-102, 1985. (On Kleeman.) 6. *Concordiensis,* Vol. 45, No. 45, March 25, 1922. 7. Union College catalogs, 1919-90. 8. "Laser Controversy," E. Larson, *Science Digest,* pages 12-23, January, 1990. 9. *Who's Who in America,* database, 1988. 10. Personal interview with David Berley, '51. 11. American Physical Society Membership Directories, 1975-90. 12. Baccalaureate Sources of Ph.D.s, Office of Institutional Research, Franklin and Marshall College, Lancaster, PA, 1993.

# NOTES

## CHAPTER 1: UNION'S FORMATIVE YEARS: 1795–1814

[1]*Rules and Regulations for the Government of the Academy in Schenectady* (Albany: Barber and Southwick, 1793).

[2]John Taylor file, Union College Archives.

[3]Samuel Miller, *Brief Retrospect of the 18th Century,* 2 vol. (New York: T and J. Swords, 1803), Vol. 2, p. 39.

[4]A. Van Vranken Raymond, *Union University,* 3 vol. (New York: Lewis Publishing Co., 1907), Vol. 1, p. 524.

[5]The 27 colleges surveyed by Miller in 1800 were:

| | | |
|---|---|---|
| Harvard | College of N. J.(Princeton) | Hampden-Sydney (Va.) |
| Williams | University of Pennsylvania | Franklin (Pa.) |
| Bowdoin | University of N. C. | Dickinson (Pa.) |
| Dartmouth | University of Georgia | Charleston (S.C.) |
| Rhode Island | St. John's (Md.) | Winnesborough (S.C.) |
| Connecticut | Washington (Md.) | Cambridge (S.C.) |
| Middlebury | Catholic (Md.) | Beaufort (S.C.) |
| Columbia | Cokesbury (Md.) | Transylvania (Ky.) |
| Union | William and Mary (Va.) | Greenville (Tenn.) |

[6]Theodore Hornberger, *Scientific Thought in the American Colleges, 1638–1800* (Austin, Texas: University of Texas Press, 1935), p. 6.

[7]Miller, Vol. 2, p. 492.

[8]John Bellamy Taylor File, Union College Archives.

[9]Wayne Somers, *Early Scientific Books in Schaffer Library* (Schenectady: Union College, 1971).

[10]*The Schenectady Cabinet,* June 8, 1824.

[11]*Union College Magazine,* Vol. X, No. 1. November, 1871.

[12]Codman Hislop, *Eliphalet Nott* (Middletown, Conn.: Wesleyan University Press, 1971), p. 49.

[13]Franklin B. Hough, *Historical Sketch of Union College* (Washington, D.C.: U.S. Government Printing Office, 1876), p. 70.

[14]*First Semi-centennial of the Philomathean Society,* July 25, 1848 (Albany: Weed Parsons & Co., 1849).

[15]Hislop, p. 224.

[16]William Enfield, *Institutes of Natural Philosophy* (London: T. Johnson, 1783, 1785, 1799; American ed. edited by Samuel Weber, Boston: 1802, 1811. 3rd. ed. Boston: Cummings and Hilliard, 1820).

[17]Stanley M. Guralnick, *Science and the Ante-Bellum American College* (Philadelphia: American Philosophical Society, 1975), p. 62.

[18]Hislop, p. 256.

# CHAPTER 2: DEVELOPING SCIENTIFIC LEADERSHIP: 1815–1826

[1]Guralnick, p. 18

[2]Hislop, p. 202.

[3]*Ibid.*, p. 226.

[4]Guralnick, p. 22.

[5]*Ibid.*, pp. 22, 23.

[6]*Ibid.,* pp. 171, 197.

[7]Hislop, p. 225.


# CHAPTER 3: UNION COLLEGE'S HEYDAY: 1827–1860

[1]Hislop, p. 220.

[2]R. F. Butts, *The College Charts Its Course* (New York: McGraw Hill, 1939), p. 136.

[3]Hislop, p. 582; L.F. Snow, *The College Curriculum in the United States* (New York: Teachers College, Columbia University, 1907), p. 158.

[4]Frederick Rudolph, *Curriculum: A History of the American Undergraduate Course of Study Since 1636* (San Francisco: Jossey-Bass, 1978), p.75; Guralnick, p. 25.

[5]Hislop, p. 219.

[6]Rudolph, p. 86.

[7]*Ibid.*, p. 85.

[8]Hislop, p. 232.

[9]*Ibid.*, p. 230.

[10]*Ibid.*, pp. 230, 231.

[11]Jonathan Pearson, *Diary*. Vol. 1, Entry of Sept. 23, 1834.

[12]Guralnick, p. 58.

[13]Joseph Henry, *The Papers of Joseph Henry*, ed. by Nathan Reingold (Washington, D.C.: Smithsonian Institution Press, 1972- ), Vol. I., p. 276, May 6, 1830.

Henry, who came from Galway, N.Y., never attended college but was recognized early for his outstanding achievements in scientific research. He received an honorary degree from Union in 1829, and later taught at Princeton. He was the first Director of the Smithsonian Institution.

[14]Asa Fitch Diary, pp. 84–89. Yale University Library.

[15]Hislop, p. 399.

[16]Leander Hall, *Union College: Half-Century History of the Class of 1856 (1906)*, p. 27.

[17]Hough, p. 71.

[18]Catalog of the Historical Exhibition, Union College, October 1934, p. 19. (Supplement to the *Union Alumni Monthly,* Vol. XXIV, No. 1.)

[19]Henry, Vol. 6, p. 320, Oct. 25, 1845.

# CHAPTER 4: THE TROUBLED YEARS: 1861-1885

[1]Hislop, pp. 549, 559.

[2]John Foster, *Open Letter to the Alumni of Union College* (Schenectady: Privately printed, 1894), p. 49.

[3]George Howell and John H. Munsell, *History of the County of Schenectady* (New York: W.W. Munsell & Co., 1886), p. 137.

[4]John Foster, *An Elementary Lecture on Electricity, Magnetism, Galvanism, Electromagnetism and Acoustics* (Schenectady: James H. Barhyte, 1877).

[5]Item in a file entitled "Reminiscences of Union College," College Archives.

[6]Frank W. Clarke, *A Report on the Teaching of Chemistry and Physics in the United States* (Washington, D.C.: Government Printing Office, 1881) p. 402.

[7]*Ibid,* p. 403.

[8]*The Concordiensis,* Vol. 4, No. 7, April 1881, pp. 91 and 102.

[9]Foster, *Open Letter,* p. 63.

[10]George E. Peterson, *The New England College in the Age of the University* (Amherst, Ma.: Amherst College Press, 1964), p. 140.


# CHAPTER 5: TRANSITION TO A NEW CENTURY: 1885-1900

[1]Richard F. Axen, "History and Analysis of a Liberal Arts College Curriculum: Four Perspectives of Union College." Unpublished dissertation, University of California, 1952, p. 110.

[2]Raymond, Vol. I, p. 407.

[3]*Ibid,* p. 432.

# SELECTED BIBLIOGRAPHY

## UNPUBLISHED SOURCES FROM UNION COLLEGE SPECIAL COLLECTIONS:

Axen, Richard F. "History and Analysis of a Liberal Arts College Curriculum: Four Perspectives of Union College." Dissertation, University of California, 1952.

Bacon, Egbert K. "The First Hundred Years of Chemistry at Union College, 1810-1910."

Catalog of Library Holdings, 1799.

Catalog of Library Holdings, 1815.

Faculty Minutes, 1799-.

Files on Individual Faculty and Alumni.

Foster, John. Diaries.

George, Carl and Uzzo, Robert. "The Nott Memorial—Symbolic Elements in the Architecture of Edward Tuckerman Potter."

Inventory of the Library and Apparatus owned by Professor William Gillespie, 1868.

Jackson, Isaac. Diaries and Notebooks.

Minutes of the Board of Trustees of Union College, 1795- .

Nott, Eliphalet. First Baccalaureate Address, May 1, 1805.

_____. Letter to his brother Samuel, April 14, 1810.

Pearson, Jonathan. Diaries and Scrapbook.

"Reminiscences of Union College." File in College Archives.

Reports to the New York State Board of Regents.

Smith, John Blair. Inaugural Address, May 1, 1795.

Student Notebooks and Diaries.

Treasurer's Books.

Union College Letter Books, ca. 1820-1870.

Union College Merit Books.

Union College Physical Laboratory, 2 vol., 1875. An illustrated inventory of more than 700 items of physical apparatus.

## OTHER UNPUBLISHED SOURCES:

Fitch, Asa. Diary. (Yale University Library)

# PUBLISHED MATERIALS:

*Annual Reports of the Regents of the University of the State of New York.*

Boss, Benjamin. *History of the Dudley Observatory, 1852–1956.* Albany: The Dudley Observatory, 1968.

Butts, R.F. *The College Charts Its Course.* New York City: McGraw Hill, 1939.

Catalog of the Historical Exhibition, Union College, October 1934. (Supplement to *Union Alumni Monthly,* Vol. XXIV, No. 1.)

Clarke, Frank W. *A Report on the Teaching of Chemistry and Physics in the United States.* Washington, D.C.: U.S. Govt. Printing Office, 1881.

*The Concordiensis.* Vol. 4, No.7. April, 1881.

Daumas, Maurice. *Scientific Instruments of the Seventeenth and Eighteenth Centuries and their Makers.* Translated and edited by Dr. Mary Holbrook. London: Portman Books, 1972.

De Clercq, P.R., ed. *Nineteenth-Century Scientific Instruments and their Makers.* Amsterdam: Rodopi, 1985.

Enfield, William. *Institutes of Natural Philosophy.* London: S. Johnson, 1785. 2nd American ed. edited by Samuel Weber. Boston: Thomas and Andrews, 1811. 3rd American ed. Boston: Cummings and Hilliard, 1820.

Farrar, John. *An Elementary Treatise on Mechanics.* Cambridge, Ma: Hilliard and Metcalf, 1825.

*First Semi-centennial Anniversary of the Philomathean Society, July 25, 1848.* Albany: Weed Parsons & Co., 1849.

*The First Semi-Centennial Anniversary of Union College.* Schenectady: I. Riggs, 1845.

Fortenbaugh, Samuel. *In Order to Form a More Perfect Union.* Schenectady: Union College Press, 1978.

Foster, John. *An Elementary Lecture on Electricity, Magnetism, Galvanism, Electromagnetism and Acoustics.* Schenectady: 1877.

_____. *An Open Letter to the Alumni of Union College.* Schenectady: 1894.

Fox, Dixon Ryan. *Union College, An Unfinished History.* Schenectady: Graduate Council, Union College, 1945.

Ganot, Adolphe. *Elementary Treatise on Physics, Experimental and Applied.* Translated by E. Atkinson. London: 1863.

Guralnick, Stanley M. *Science and the Ante-Bellum American College.* Philadelphia: American Philosophical Society, 1975.

Hall, Leander. *Union College: Half-Century History of the Class of 1856.* 1906.

Henry, Joseph. *The Papers of Joseph Henry.* Edited by Nathan Reingold. Washington, D.C.: The Smithsonian Institution Press, 1972- .

Hislop, Codman. *Eliphalet Nott.* Middletown, Conn.: Wesleyan University Press, 1971.

Hornberger, Theodore. *Scientific Thought in the American Colleges, 1638–1800.* Austin, Texas: University of Texas Press, 1935.

Hough, Franklin B. *Historical Sketch of Union College.* Washington, D.C.: Government Printing Office, 1876.

Howell, George and Munsell, John H. *History of the County of Schenectady, New York.* Schenectady: W.W. Munsell and Co., 1886.

Jackson, Isaac. *Elementary Treatise on Optics.* New York: 1848.

_____. *Elementary Treatise on Mechanics.* Albany: 1852.

_____. *Elements of Conic Sections.* Schenectady: 1836.

_____. *Elements of Trigonometry.* Schenectady: 1859.

Joslin, Benjamin. *Meteorological Observations and Essays.* Albany: Packard and Van Benthuysen, 1836.

Lardner, Dionysus. *A Treatise on Hydrostatics and Pneumatics.* First American edition edited by Benjamin F. Joslin. Philadelphia: Cary and Lea, 1832.

Miller, Samuel. *A Brief Retrospect of the 18th Century*, 2 vol. New York: T. and J. Swords, 1803.

Mills, John F. *Encyclopedia of Antique Scientific Instruments.* London: Aurum Press, 1983.

Peterson, George E. *The New England College in the Age of the University.* Amherst, Ma.: Amherst College Press, 1964.

*Pike's Illustrated Catalogue of Scientific and Medical Instruments.* A facsimile edition. Dracut, Mass.: The Antiquarian Scientist, 1984.

Potter, Alonzo. *The Principles of Science, Applied to Domestic and Mechanic Arts.* Boston: Marsh, Capen, Lyon, & Webb, 1841.

Raymond, A. Van Vranken. *Union University,* 3 Vol. New York: Lewis Publishing Co., 1907.

*Rittenhouse: Journal of the American Scientific Enterprise.* Dracut, Ma.: The Antiquarian Scientist, 1986- .

Rudolph, Frederick. *Curriculum: A History of the American Undergraduate Course of Study Since 1636.* San Francisco: Jossey-Bass, 1978.

*Rules and Regulations for the Government of the Academy in Schenectady.* Albany: Barber and Southwick, 1793.

*Semi-Centennial Anniversary of the Connection of Professor Isaac Jackson with Union College.* Schenectady: 1876.

Somers, Wayne. *Early Scientific Books in Schaffer Library, Union College.* Schenectady: Union College, 1971.

_____. *Perseverance Conquers Much.* Schenectady: Friends of the Union College Library, 1990.

Stone, William. *The Olivier Models.* Schenectady: Friends of the Union College Library, 1969.

Turner, Gerard L'e. *Antique Scientific Instruments.* Poole: Blandford Press, 1980.

_____. *Nineteenth-Century Scientific Instruments.* Berkeley: University of California Press, 1983.

Union College Catalogs, 1805- .

Union College Laws and Regulations, 1795- .

Union College Alumni Magazines.

Wheatland, David P. *The Apparatus of Science at Harvard, 1765-1800.* Cambridge: Harvard University Press, 1968.

# INDEX